AF793952

Springer Monographs in Mathematics

Springer-Verlag London Ltd.

Raymond A. Ryan

Introduction to Tensor Products of Banach Spaces

With 31 Figures

Springer

Raymond A. Ryan
Department of Mathematics
National University of Ireland, Galway
Galway
Ireland

British Library Cataloguing in Publication Data
Ryan, Raymond A.
Introduction to tensor products of Banach spaces. -
(Springer monographs in mathematics)
1. Banach spaces 2. Tensor products
I. Title
515.7'32
ISBN 978-1-84996-872-0
Library of Congress Cataloging-in-Publication Data
Ryan, Raymond A.
Introduction to tensor products of Banach spaces / Raymond A. Ryan.
p. cm. -- (Springer monographs in mathematics)
Includes bibliographical references and index.
ISBN 978-1-84996-872-0 ISBN 978-1-4471-3903-4 (eBook)
DOI 10.1007/978-1-4471-3903-4
1. Banach spaces 2. Tensor products. I. Title. II. Series.
QA322.2.R92 2002
515'.732—dc21 2001049821

Mathematics Subject Classification (2000): 46-01, 47-01, 46A32, 46B22, 46B28, 46M05, 47B10

Apart from any fair dealing for the purposes of research or private study, or criticism or review, as permitted under the Copyright, Designs and Patents Act 1988, this publication may only be reproduced, stored or transmitted, in any form or by any means, with the prior permission in writing of the publishers, or in the case of reprographic reproduction in accordance with the terms of licences issued by the Copyright Licensing Agency. Enquiries concerning reproduction outside those terms should be sent to the publishers.

Springer Monographs in Mathematics ISSN 1439-7382
ISBN 978-1-84996-872-0

http://www.springer.co.uk

© Springer-Verlag London 2002
Originally published by Springer-Verlag London Berlin Heidelberg 2002

The use of registered names, trademarks, etc. in this publication does not imply, even in the absence of a specific statement, that such names are exempt from the relevant laws and regulations and therefore free for general use.

The publisher makes no representation, express or implied, with regard to the accuracy of the information contained in this book and cannot accept any legal responsibility or liability for any errors or omissions that may be made.

Typesetting: Camera-ready by the author

12/3830-543210 Printed on acid-free paper SPIN 10792170

To my parents

To my parents

Preface

This book is intended as an introduction to the theory of tensor products of Banach spaces. The prerequisites for reading the book are a first course in Functional Analysis and in Measure Theory, as far as the Radon–Nikodým theorem. The book is entirely self-contained and two appendices give additional material on Banach Spaces and Measure Theory that may be unfamiliar to the beginner. No knowledge of tensor products is assumed.

Our viewpoint is that tensor products are a natural and productive way to understand many of the themes of modern Banach space theory and that "tensorial thinking" yields insights into many otherwise mysterious phenomena. We hope to convince the reader of the validity of this belief.

We begin in Chapter 1 with a treatment of the purely algebraic theory of tensor products of vector spaces. We emphasize the use of the tensor product as a linearizing tool and we explain the use of tensor products in the duality theory of spaces of operators in finite dimensions. The ideas developed here, though simple, are fundamental for the rest of the book.

In Chapter 2 we introduce the projective tensor product. The space of bounded bilinear forms on the product $X \times Y$ of two Banach spaces is the dual of the projective tensor product, $X\hat{\otimes}_\pi Y$. Projective tensor products with ℓ_1 and $L_1(\mu)$ are particularly easy to interpret and the Bochner integral for vector valued functions is introduced. The first hint of the influence of finite dimensional structure appears here with the definition of the $\mathcal{L}_1$-spaces. We give some elementary applications of the Rademacher functions, including the Khinchine Inequality and its implications for the structure of tensor products of $L_p(\mu)$ spaces. Finally, we introduce the class of nuclear operators.

The injective tensor product is studied in Chapter 3, along with the class of integral operators. The attempt to represent the injective tensor product $L_1(\mu)\hat{\otimes}_\varepsilon X$ as a space of vector valued functions leads to the definition of the Pettis integral and among the applications we prove the Orlicz–Pettis theorem, which provides a bridge between weak and norm summability. The importance of the class of integral operators derives in part from the fact that certain canonical operators, such as the injection of $L_\infty(\mu)$ into $L_1(\mu)$, where μ is a finite measure, are of this type. We study this idea in detail.

Chapter 4 is devoted to the approximation property and related matters. The chapter begins with a list of open problems from the previous chapters, all of which depend on the presence of the approximation property in some form for their solution. We give a detailed account of the property and its applications. We then turn to the question of when the projective, or injective, tensor product of reflexive spaces is also reflexive. The chapter concludes with an examination of the construction of tensor product bases for spaces that have Schauder bases.

The Radon–Nikodým property, and its role in the theory of tensor products, is the subject of Chapter 5. We give a complete account of the vector measure theory needed for this purpose. Projective and injective tensor products with the space of measures are characterized in terms of spaces of vector measures. We develop the representation theory of operators on $C(K)$ spaces by means of vector measures, culminating in the phenomenon of the coincidence of the classes of integral and nuclear operators in the presence of the Radon–Nikodým property. The possession of this property by reflexive Banach spaces and separable duals, among others, is established through a reformulation of the property in terms of the representability of operators on $L_1(\mu)$ spaces. We present several striking applications that illustrate the power of the Radon–Nikodým property, finishing with the Principle of Local Reflexivity.

Chapter 6 begins with a discussion of uniform crossnorms and the central concept of a tensor norm. The Chevet–Saphar norms g_p and d_p are introduced, leading to a study of the class of p-summing operators. Among the results treated here are the Pietsch Domination Theorem for p-summing operators. The highlight of the chapter is the Grothendieck inequality. After proving this fundamental result, we give some of its applications to operators between classical Banach spaces.

In Chapter 7 we embark on a thorough investigation of the basic properties of tensor norms. We begin with Schatten's construction of a dual norm, pointing out its shortcomings and the benefits to be obtained from adopting the Grothendieck definition. The question of when the two duals coincide leads in a natural way to the concept of accessibility for a tensor norm. After examining the interconnections between accessibility, duality and the possession of injective or projective properties, we study the injective and projective associate norms that can be generated from a tensor norm. This machinery enables us to identify the duals of the Chevet–Saphar tensor norms and this in turn leads to the classes of p-integral operators. We then turn our attention to the Hilbertian tensor norm, w_2, and the central role it plays in the theory by virtue of the Grothendieck Inequality. After an account of the associated classes of 2-factorable, or Hilbertian, and 2-dominated operators, we conclude with Grothendieck's classification of the fourteen "natural" tensor norms.

In Chapter 8 we present a very brief introduction to the theory of Operator Ideals, concentrating on the connection with the parallel theory of Tensor Norms. Each tensor norm α generates two Banach ideals, the ideal of α-nuclear operators and the ideal of α-integral operators, which coincide in finite dimensions. We show that these ideals are the smallest and largest ideals respectively that are associated with α. We also look at the relationship between a Banach ideal and the tensor norm associated with it.

Each chapter is accompanied by a set of exercises. They are, for the most part, straightforward applications of the subject matter of the chapter in question, and are designed to help the reader in coming to terms with the concepts introduced there. However, we have resisted the temptation to use the exercises as a device for introducing topics that could not be accommodated in the text for reasons of space.

We make no claims whatever to originality, except perhaps in the arrangement of the material. Our overriding objective has been to provide a self-contained, accessible and compact introduction to the subject. We have been highly selective in our choice of topics, preferring to spend time on a small set of important themes rather than include everything one might like to have seen. The expert will find that many of his or her favourite areas have been omitted. We hope that the reader will go on to study the more comprehensive works mentioned in Appendix A. Most of all, we hope that the reader will enjoy this book.

It is a great pleasure to thank all those who have helped me in one way or another. My colleagues in the Mathematics Department in Galway have always been supportive. Departmental heads Ted Hurley and Martin Newell provided crucial practical assistance. My students, Pádraig Kirwan and Bogdan Grecu, learned much of this material with me and read many early drafts. I am grateful to Raymundo Alencar, Richard Aron, Juan Bes, Fernando Blasco, Joe Diestel, Klaus Floret, Mikael Lindström, Manuel Maestre, Yannis Sarantopoulos, Andrew Tonge, Barry Turret and Ignacio Zalduendo for many helpful conversations and for sharing their knowledge with me. Seán Dineen provided vital encouragement and advice at critical moments. Michael Mackey, Niall Madden, Götz Pfeiffer and Dirk Werner all gave invaluable assistance with TEX-related matters. Jose Ansemil and Chris Boyd gave generously of their time in reading drafts of the book and providing many corrections, suggestions and ideas. Dirk Werner not only read and corrected many drafts, but was also a knowledgeable and attentive sounding board. Without his deep knowledge and experience, this would have been a much smaller book.

The support of the Millennium Research Fund of the National University of Ireland, Galway, is gratefully acknowledged. The book was typeset using LATEX and the diagrams were produced using Paul Taylor's Commutative Diagrams package. Finally, I would like to express my appreciation to Karen Borthwick and all the editorial staff at the London office of Springer Verlag for their patient and professional support.

Galway, September 2001 *Ray Ryan*

Notation and Terminology

The letters X, Y, Z will denote Banach spaces over the scalar field $\mathbb{K}$, which may be either the real numbers, $\mathbb{R}$, or the complex numbers, $\mathbb{C}$. The letters E, F, M, N will usually denote finite dimensional Banach spaces. The closed unit ball of a Banach space X will be denoted by B_X.

The term *operator* will mean a bounded linear mapping. The space of operators from X into Y will be denoted by $\mathcal{L}(X, Y)$ and the dual space of X by X^*. Elements of a dual space X^* will typically be denoted by φ or ψ and elements of a bidual X^{**} by symbols such as x^{**} or y^{**}. If φ is a linear functional on X, we shall denote the value of φ at an element x of X either by $\varphi(x)$ or $\langle x, \varphi \rangle$.

We shall denote the space of bounded bilinear forms on the product $X \times Y$ of two Banach spaces by $\mathcal{B}(X \times Y)$, with the norm given by $\|B\| = \sup\{|B(x,y)| : x \in B_X, y \in B_Y\}$.

$C(K)$ and $L_p(\mu)$ will denote the usual sequence spaces, where $1 \leq p \leq \infty$. We shall follow the traditional abuse of notation in using a symbol such as f to denote both the equivalence class of a function in $L_p(\mu)$ and the function itself.

The symbols c_0, ℓ_p, will denote the usual sequence spaces and ℓ_p^n will denote the space $\mathbb{K}^n$ with the ℓ_p-norm. If X is a Banach space, the space $\ell_p(X)$ consists of all sequences $x = (x_n)$ in X for which the scalar sequence $(\|x_n\|)$ belongs to ℓ_p, with the norm of x defined to be the ℓ_p-norm of $(\|x_n\|)$. A similar remark applies to the space $c_0(X)$. The scalar sequence spaces obtained when the indexing set $\mathbb{N}$ is replaced by a set I will be denoted by $c_0(I)$ and $\ell_p(I)$; the context will remove any possible notational confusion with the above-mentioned vector valued sequence spaces.

Contents

1. Tensor Products

In this chapter we study tensor products from a purely algebraic viewpoint. Our approach is to define tensors as functionals that act on bilinear forms. We explain how the tensor product can be seen as a "linearizing space" for bilinear mappings. Tensors can also be viewed as bilinear forms, or as linear mappings and we explore the connections between these ideas. In finite dimensions, tensor products provide a means of understanding the duality of spaces of linear mappings or bilinear forms, either through "tensor duality" or the equivalent "trace duality". Finally, we look at some examples of tensor products.

We work with vector spaces over the field $\mathbb{K}$, which can be either the real numbers, $\mathbb{R}$, or the complex numbers, $\mathbb{C}$. We denote the algebraic dual of the vector space X by $X^\sharp$. Thus the space $X^\sharp$ consists of all the linear functionals on X, that is, linear mappings from X into $\mathbb{K}$. If x is an element of X and φ is a linear functional on X, we shall denote the value of φ at x either by $\varphi(x)$ or by $\langle x, \varphi \rangle$. The vector space of linear mappings from X into Y is denoted by $L(X,Y)$ and of course $L(X,\mathbb{K})$ is just the space $X^\sharp$.

1.1 Tensor Products of Vector Spaces

We recall that a mapping A from the cartesian product $X \times Y$ of vector spaces into the vector space Z is *bilinear* if it is linear in each variable, that is,

$$\begin{aligned} A(\alpha_1 x_1 + \alpha_2 x_2, y) &= \alpha_1 A(x_1, y) + \alpha_2 A(x_2, y) \quad \text{and} \\ A(x, \beta_1 y_1 + \beta_2 y_2) &= \beta_1 A(x, y_1) + \beta_2 A(x, y_2) \end{aligned}$$

for all $x_i, x \in X$, $y_i, y \in Y$ and all scalars α_i, β_i, $i = 1, 2$. We write $B(X \times Y, Z)$ for the vector space of bilinear mappings from the product $X \times Y$ into Z; when Z is the scalar field we denote the corresponding space of *bilinear forms* simply by $B(X \times Y)$.

Now the *tensor product*, $X \otimes Y$, of the vector spaces X, Y can be constructed as a space of linear functionals on $B(X \times Y)$, in the following way: for $x \in X$, $y \in Y$, we denote by $x \otimes y$ the functional given by evaluation at the point (x, y). In other words,

$$(x \otimes y)(A) = \langle A, x \otimes y \rangle = A(x, y)$$

for each bilinear form A on $X \times Y$. The tensor product $X \otimes Y$ is the subspace of the dual $B(X \times Y)^\sharp$ spanned by these elements. Thus, a typical *tensor* in $X \otimes Y$ has the form

$$u = \sum_{i=1}^{n} \lambda_i \, x_i \otimes y_i \,, \tag{1.1}$$

where n is a natural number, $\lambda_i \in \mathbb{K}$, $x_i \in X$ and $y_i \in Y$. It is important to realize that the representation of u is not unique – in general, there will be many different ways to write a given tensor in the above form. We shall deal with this question later. For now, we note a few elementary facts about tensors. First, if $u = \sum_{i=1}^{n} \lambda_i \, x_i \otimes y_i$ is a tensor and A a bilinear form, then the action of u on A is given by

$$u(A) = \left\langle A, \sum_{i=1}^{n} \lambda_i \, x_i \otimes y_i \right\rangle = \sum_{i=1}^{n} \lambda_i \, A(x_i, y_i) \,.$$

Of course, the value of this expression is independent of the particular representation chosen for u. Next, we observe that we can think of the mapping $(x, y) \mapsto x \otimes y$ as a sort of multiplication on $X \times Y$ with values in the vector space $X \otimes Y$. This "product" is itself bilinear, so we have, for example,

(i) $(x_1 + x_2) \otimes y = x_1 \otimes y + x_2 \otimes y$,
(ii) $x \otimes (y_1 + y_2) = x \otimes y_1 + x \otimes y_2$,
(iii) $\lambda(x \otimes y) = (\lambda x) \otimes y = x \otimes (\lambda y)$,
(iv) $0 \otimes y = x \otimes 0 = 0$.

It is worth noting that the typical description of an element u of $X \otimes Y$ given in (1.1) can always be rewritten, using (iii) above, in the form

$$u = \sum_{i=1}^{n} x_i \otimes y_i \,.$$

Our first proposition shows how linearly independent sets and bases can be transferred from the component spaces into the tensor product.

Proposition 1.1. *Let X and Y be vector spaces.*

(a) *Let E and F be linearly independent subsets of X and Y respectively. Then $\{x \otimes y : x \in E,\ y \in F\}$ is a linearly independent subset of $X \otimes Y$.*
(b) *If $\{e_i : i \in I\}$ and $\{f_j : j \in J\}$ are bases for X, Y respectively then $\{e_i \otimes f_j : (i, j) \in I \times J\}$ is a basis for $X \otimes Y$.*

Proof. We prove (a), from which (b) follows immediately. Suppose we have $u = \sum_{i=1}^{n} \lambda_i \, x_i \otimes y_i = 0$, where $x_i \in E$ and $y_i \in F$. Let φ, ψ be linear functionals on X, Y respectively and consider the bilinear form defined by $A(x,y) = \varphi(x)\psi(y)$. We have $u(A) = 0$ and so

$$\sum_{i=1}^{n} \lambda_i \varphi(x_i)\psi(y_i) = \psi\left(\sum_{i=1}^{n} \lambda_i \varphi(x_i) y_i\right) = 0\,.$$

Since this holds for every $\psi \in Y^\sharp$, we can conclude that $\sum_{i=1}^{n} \lambda_i \varphi(x_i) y_i = 0$ and so, by the linear independence of F, we have $\lambda_i \varphi(x_i) = 0$ for every $\varphi \in X^\sharp$. But, by the linear independence of E, each x_i is nonzero and it follows that $\lambda_i = 0$ for every i. □

It follows from this result that if X and Y are finite dimensional spaces, then

$$\dim(X \otimes Y) = \dim(X)\dim(Y)\,.$$

In particular, we see that, taking one of the component spaces to be the scalar field, the space $X \otimes \mathbb{K}$ has the same dimension as X. In fact, regardless of dimensional considerations, the tensor products $X \otimes \mathbb{K}$ and $\mathbb{K} \otimes X$ are each canonically isomorphic to X; we leave it as an exercise to the reader to show that the identifications $x \otimes \lambda \mapsto \lambda x$ and $\lambda \otimes x \mapsto \lambda x$ respectively provide the required isomorphisms.

For every non-zero tensor $u \in X \otimes Y$ there is a smallest natural number n for which there is a representation of u containing n terms; let $\sum_{i=1}^{n} x_i \otimes y_i$ be such a representation. Then it is clear that the sets $\{x_1, \ldots, x_n\}$ and $\{y_1, \ldots, y_n\}$ are each linearly independent. Suppose, for example, that x_1 were a linear combination of $x_2, \ldots, x_n$; then the term $x_1 \otimes y_1$ could be absorbed by the others, giving a representation of u with $n-1$ terms. The number n is known as the *rank* of u. Tensors of rank one are often referred to as *elementary tensors*.

How can we tell if two tensors are the same? This question reduces to the following: how is it possible to determine whether $\sum_{i=1}^{n} x_i \otimes y_i$ is a representation of the zero tensor? In principle, this can be determined by evaluating $\sum_{i=1}^{n} A(x_i, y_i)$ for every bilinear form A. Fortunately, there are easier ways to proceed.

Proposition 1.2. *The following are equivalent for* $u = \sum_{i=1}^{n} x_i \otimes y_i \in X \otimes Y$*:*

(i) $u = 0$;
(ii) $\sum_{i=1}^{n} \varphi(x_i)\psi(y_i) = 0$ *for every* $\varphi \in X^\sharp, \psi \in Y^\sharp$;
(iii) $\sum_{i=1}^{n} \varphi(x_i) y_i = 0$ *for every* $\varphi \in X^\sharp$.
(iv) $\sum_{i=1}^{n} \psi(y_i) x_i = 0$ *for every* $\psi \in Y^\sharp$.

Proof. The proofs of (i) ⇒ (ii) ⇒ (iii) ⇒ (iv) are straightforward variations of the proof of Proposition 1.1 and are left to the reader. To prove (iv) ⇒ (i), suppose that $\sum_{i=1}^{n} \psi(y_i) x_i = 0$ for every $\psi \in Y^\sharp$. Let $A \in B(X \times Y)$. Let E, F be the subspaces of X, Y spanned by $\{x_1, \ldots, x_n\}$, $\{y_1, \ldots, y_n\}$ respectively and let B denote the restriction of A to $E \times F$. Choosing bases for the finite dimensional spaces E, F and expanding the bilinear form B relative to these bases yields a representation for B of the form

$$B(x,y) = \sum_{j=1}^{m} \theta_j(x)\omega_j(y)$$

where $\theta_j \in E^\sharp$ and $\omega_j \in F^\sharp$. We may extend the domains of θ_j, ω_j to all of X, Y respectively in the following manner: choose algebraic complements, G, H for E, F respectively, so that $X = E \oplus G$ and $Y = F \oplus H$. Then, if $x = x_1 + x_2 \in X$ with $x_1 \in E, x_2 \in G$, let $\theta_j(x) = \theta_j(x_1)$. The functionals ω_j are defined on Y in a similar way. We may now consider B as a bilinear form on $X \times Y$ by using the representation of B given above. Now A and B may be different bilinear forms on $X \times Y$, but they agree on $E \times F$. Thus we have

$$\begin{aligned} u(A) &= \sum_{i=1}^{n} A(x_i, y_i) = \sum_{i=1}^{n} B(x_i, y_i) \\ &= \sum_{i=1}^{n}\sum_{j=1}^{m} \theta_j(x_i)\omega_j(y_i) = \sum_{j=1}^{m} \theta_j\left(\sum_{i=1}^{n} \omega_j(y_i)x_i\right) = 0\,, \end{aligned}$$

using (iv). Thus $u(A) = 0$ for every $A \in B(X \times Y)$. □

This proposition provides effective means of recognizing when two tensors are the same and we will make extensive use of it. A subset, S, of a dual space, $X^\sharp$, is said to be *separating* if it contains enough functionals to distinguish between the points of X; in other words, if $\varphi(x) = 0$ for every $\varphi \in S$, then x must be zero. It is easy to see that, in applying the proposition, the functionals used may be restricted to lie in separating subsets of the relevant duals. An important instance of this idea occurs when the component spaces in the tensor product are dual spaces: in order to show that $\sum_{i=1}^{n} \varphi_i \otimes \psi_i = 0$ in $X^\sharp \otimes Y^\sharp$, it suffices to have $\sum_{i=1}^{n} \varphi_i(x)\psi_i(y) = 0$ for every $x \in X, y \in Y$. We have similar variations on statements (ii) and (iii) of the proposition.

Next, we look at the interaction between tensor products and some commonly used vector space constructions. Suppose that E and F are subspaces of the vector spaces X and Y respectively. Then $E \otimes F$ can be considered as a subspace of the tensor product $X \otimes Y$ in a natural way – if $u = \sum_{i=1}^{n} x_i \otimes y_i$ is an element of $E \otimes F$, we can think of u as an element of $X \otimes Y$ by the formula $\langle A, u\rangle = \sum_{i=1}^{n} A(x_i, y_i)$, where $A \in B(X \times Y)$. This gives us a linear mapping of $E \otimes F$ into $X \otimes Y$. To see that this mapping is injective, suppose that $\sum_{i=1}^{n} x_i \otimes y_i = 0$ in $X \otimes Y$. We have seen in the proof of the above proposition that for every bilinear form B on $E \times F$ there is a bilinear form A on $X \times Y$ such that $\sum_{i=1}^{n} B(x_i, y_i) = \sum_{i=1}^{n} A(x_i, y_i)$. Therefore $\sum_{i=1}^{n} B(x_i, y_i) = 0$ and it follows that $\sum_{i=1}^{n} x_i \otimes y_i = 0$ in $E \otimes F$.

Next, we show that tensor products respect direct sums:

Proposition 1.3. *If* $Y = F_0 \oplus F_1$ *then* $X \otimes Y = (X \otimes F_0) \oplus (X \otimes F_1)$.

Proof. It is clear that $X \otimes Y$ is spanned by the subspaces $X \otimes F_0$ and $X \otimes F_1$. It only remains to show that $(X \otimes F_0) \cap (X \otimes F_1) = \{0\}$. Let $u \in (X \otimes F_0) \cap (X \otimes F_1)$, so that we have two representations of u:

$$u = \sum_{i=1}^{n} v_i \otimes y_i = \sum_{j=1}^{m} w_j \otimes z_j \,,$$

with $y_i \in F_0$ and $z_j \in F_1$. Then for every $\varphi \in X^\sharp$ we have $\sum_{i=1}^n \varphi(v_i)y_i = \sum_{j=1}^m \varphi(w_j)z_j$. But $F_0 \cap F_1 = \{0\}$, and so $\sum_{i=1}^n \varphi(v_i)y_i = 0$ for every $\varphi \in X^\sharp$. It follows that $u = 0$. □

It follows from this result that the quotient space $(X \otimes Y)/(X \otimes F_0)$ can be identified with $X \otimes (Y/F_0)$. Of course, the same assertions hold for subspaces and quotients of the first component of a tensor product.

1.2 Tensor Products and Linearization

The primary purpose of tensor products is to linearize bilinear mappings. To explain this, let $A\colon X \times Y \to \mathbb{K}$ be a bilinear form. We recall that each tensor $u \in X \otimes Y$ acts as a linear functional on the space of bilinear forms and so we may define a mapping $\tilde{A}\colon X \otimes Y \to \mathbb{K}$ by $u \in X \otimes Y \mapsto \langle A, u\rangle \in \mathbb{K}$. Now $\tilde{A}$ is easily seen to be a linear functional on $X \otimes Y$. Furthermore, we have $\tilde{A}(x \otimes y) = \langle A, x \otimes y\rangle = A(x, y)$ and it follows that

$$\tilde{A}\Big(\sum_{i=1}^{n} x_i \otimes y_i\Big) = \sum_{i=1}^{n} A(x_i, y_i)\,.$$

We have factored the bilinear form A as the composition of the special bilinear mapping $(x, y) \in X \times Y \mapsto x \otimes y \in X \otimes Y$ and the linear functional $\tilde{A}$ that maps $X \otimes Y$ into the scalar field. On the other hand, it is easy to see that if Ψ is a linear functional on the tensor product $X \otimes Y$, then the composition of Ψ with the bilinear mapping $(x, y) \mapsto x \otimes y$ is a bilinear form, A, on $X \times Y$ for which $\tilde{A} = \Psi$. Thus the bilinear forms on $X \times Y$ are in one-to-one correspondence with the linear functionals on $X \otimes Y$. In summary, we have

$$B(X \times Y) = (X \otimes Y)^\sharp\,.$$

The same idea can be applied to bilinear mappings. If $A\colon X \times Y \to Z$ is a bilinear mapping, we define a linear mapping $\tilde{A}\colon X \otimes Y \to Z$ by $\tilde{A}(\sum_{i=1}^n x_i \otimes y_i) = \sum_{i=1}^n A(x_i, y_i)$. To show that this mapping is well defined, suppose that $\sum_{i=1}^n x_i \otimes y_i = 0$. Then, for each $\varphi \in Z^\sharp$, the composition $\varphi \circ A$ is a bilinear functional on $X \times Y$ and so

$$\varphi\Big(\sum_{i=1}^{n} A(x_i, y_i)\Big) = \sum_{i=1}^{n} \varphi \circ A(x_i, y_i) = \Big\langle \sum_{i=1}^{n} x_i \otimes y_i, \varphi \circ A\Big\rangle = 0\,.$$

It follows that $\sum_{i=1}^{n} A(x_i, y_i) = 0$. Therefore $\tilde{A}$ is well defined. Thus the bilinear mapping A is associated with the linear mapping $\tilde{A}$ and, as above, this yields an identification:

$$B(X \times Y, Z) = L(X \otimes Y, Z).$$

The situation is illustrated in the diagram below:

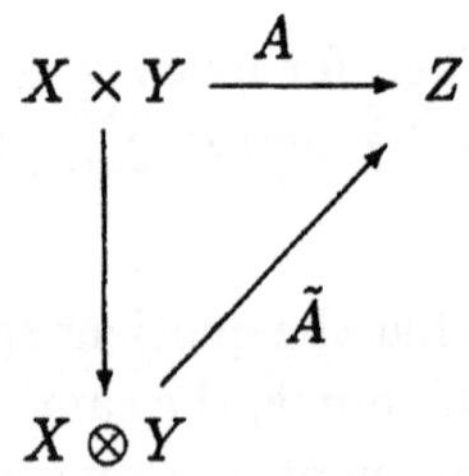

The special bilinear mapping $(x, y) \in X \times Y \mapsto x \otimes y \in X \otimes Y$ acts as a "universal" bilinear mapping – every other bilinear mapping on $X \times Y$ factors through this one via a linear mapping on the tensor product. We summarize these results:

Proposition 1.4. *For every bilinear mapping $A\colon X \times Y \to Z$ there exists a unique linear mapping $\tilde{A}\colon X \otimes Y \to Z$ such that $A(x, y) = \tilde{A}(x \otimes y)$ for all $x \in X, y \in Y$. The correspondence $A \longleftrightarrow \tilde{A}$ is an isomorphism between the vector spaces $B(X \times Y, Z)$ and $L(X \otimes Y, Z)$.*

Now, we have chosen to define the tensor product $X \otimes Y$ in a particular way, as a space of linear functionals on $B(X \times Y)$. There are other ways to construct the tensor product, some of which we shall meet in the next section. The next result demonstrates that these approaches yield essentially the same object.

Proposition 1.5 (Uniqueness of the Tensor Product). *Let X and Y be vector spaces. Suppose there exists a vector space W and a bilinear mapping $B\colon X \times Y \to W$ with the property that, for every vector space Z and every bilinear mapping A from $X \times Y$ into Z, there is a unique linear mapping $L\colon W \to Z$ such that $A = L \circ B$. Then there is an isomorphism J from $X \otimes Y$ into W such that $J(x \otimes y) = B(x, y)$ for every $x \in X$, $y \in Y$.*

Proof. First, we observe that the uniqueness of the linear mapping L for each A implies that the image of $X \times Y$ in W under B must span W. Now, applying the given property of W and B to the bilinear mapping $(x, y) \in X \times Y \mapsto x \otimes y \in X \otimes Y$, we obtain a linear mapping $L\colon W \to X \otimes Y$ such that $L(B(x, y)) = x \otimes y$ for all x, y. On the other hand, the bilinear mapping B factors through $X \otimes Y$ to give a linear mapping $\tilde{B}\colon X \otimes Y \to W$ with the property that $\tilde{B}(x \otimes y) = B(x, y)$ for every x, y. Thus, we have $\tilde{B} \circ L(B(x, y)) = B(x, y)$ and $L \circ \tilde{B}(x \otimes y) = x \otimes y$ for all x, y. Since the spaces $X \otimes Y$ and W are spanned by the elements $x \otimes y$ and $B(x, y)$ respectively, it follows that $J = \tilde{B}$ is the required isomorphism. □

We illustrate the principle that linear mappings on $X \otimes Y$ are essentially the same as bilinear mappings on $X \times Y$ with some applications. First, we show that the tensor products $X \otimes Y$ and $Y \otimes X$ are canonically isomorphic. For each $u = \sum_{i=1}^n x_i \otimes y_i \in X \otimes Y$, the *transpose* of u is the tensor

$$u^t = \left(\sum_{i=1}^n x_i \otimes y_i\right)^t = \sum_{i=1}^n y_i \otimes x_i \in Y \otimes X .$$

To see that $u \mapsto u^t$ is a well defined linear mapping, we observe that this mapping is simply the linearization of the bilinear mapping from $X \times Y$ into $Y \otimes X$ that takes (x, y) to $y \otimes x$. It is a simple matter to show that the transpose yields an isomorphism of $X \otimes Y$ with $Y \otimes X$.

We can use the same idea to define the tensor product of linear mappings. Let $S: X \to E$ and $T: Y \to F$ be linear mappings. Then we may define a bilinear mapping by $(x, y) \in X \times Y \mapsto (Sx) \otimes (Ty) \in E \otimes F$. Linearization gives a linear mapping $S \otimes T: X \otimes Y \to E \otimes F$ such that

$$(S \otimes T)(x \otimes y) = (Sx) \otimes (Ty)$$

for every $x \in X, y \in Y$. The tensor product mapping $S \otimes T$ can inherit some properties from its components. For example, we leave it to the reader to verify that if S and T are both injective (respectively, surjective) then $S \otimes T$ is also injective (respectively, surjective).

1.3 Tensors as Linear Mappings or Bilinear Forms

We have chosen to define the tensor product $X \otimes Y$ as a space of linear functionals on the space $B(X \times Y)$. There are other, equally natural approaches, some of which we describe in this section. First we show that tensors can be viewed as bilinear forms. With each pair $x \in X$, $y \in Y$ we can associate a bilinear form, $B_{x,y}$ on $X^\sharp \times Y^\sharp$, where $B_{x,y}(\varphi, \psi) = \varphi(x)\psi(y)$. The mapping $(x, y) \in X \times Y \mapsto B_{x,y} \in B(X^\sharp \times Y^\sharp)$ is easily seen to be bilinear and so there is a unique linear mapping from $X \otimes Y$ into $B(X^\sharp \times Y^\sharp)$ that maps $x \otimes y$ to $B_{x,y}$. To see that this mapping is injective, suppose that $\sum_{i=1}^n B_{x_i,y_i} = 0$. Then $\sum_{i=1}^n \varphi(x_i)\psi(y_i) = 0$ for every $\varphi \in X^\sharp, \psi \in Y^\sharp$ and so, by Proposition 1.2, it follows that $\sum_{i=1}^n x_i \otimes y_i = 0$ in $X \otimes Y$. Thus, we have a canonical embedding:

$$X \otimes Y \subset B(X^\sharp \times Y^\sharp) .$$

When the spaces in question are duals, there is a simpler embedding:

$$X^\sharp \otimes Y^\sharp \subset B(X \times Y) ,$$

where the tensor $\sum_{i=1}^n \varphi_i \otimes \psi_i$ is identified with the bilinear form that maps (x, y) to $\sum_{i=1}^n \varphi_i(x)\psi_i(y)$. In view of these canonical embeddings, we shall feel free to consider tensors as bilinear forms, should the occasion demand. If it is necessary to preserve the formal distinction between a tensor u and the bilinear form associated with it, we shall denote the bilinear form by B_u.

It is also possible to view tensors as linear mappings. Indeed, if B is a bilinear form on $X \times Y$, then we may associate two linear mappings with B – the mapping $L_B \colon X \to Y^\sharp$, and the mapping $R_B \colon Y \to X^\sharp$, defined by

$$\langle y, L_B(x)\rangle = \langle x, R_B(y)\rangle = B(x,y)\,.$$

Thus, each tensor $u = \sum_{i=1}^n x_i \otimes y_i$ generates two linear mappings, which we write as L_u and R_u:

$$L_u(\varphi) = \sum_{i=1}^n \varphi(x_i)y_i \quad \text{and} \quad R_u(\psi) = \sum_{i=1}^n \psi(y_i)x_i\,,$$

and so we have two more identifications:

$$X \otimes Y \subset L(X^\sharp, Y) \quad \text{and} \quad X \otimes Y \subset L(Y^\sharp, X)\,.$$

Finally, when one of the component spaces is a dual space, we have a natural embedding into a smaller space of linear mappings:

$$X^\sharp \otimes Y \subset L(X,Y) \quad \text{and} \quad X \otimes Y^\sharp \subset L(Y,X)\,.$$

The elements of $L(X,Y)$ that correspond to tensors in $X^\sharp \otimes Y$ under this identification are the linear mappings of finite rank, that is, mappings whose range is a finite dimensional subspace of Y.

Consider now the linear functional on the tensor product $X^\sharp \otimes X$ that linearizes the bilinear functional $(\varphi, x) \mapsto \varphi(x)$. This functional is known as the *trace* and denoted by tr, or by tr_X if it is necessary to specify the space X in question. We have

$$\mathrm{tr}\left(\sum_{i=1}^n \varphi_i \otimes x_i\right) = \sum_{i=1}^n \varphi_i(x_i)$$

and, of course, this value is independent of the representation. The trace functional is of particular interest when the tensors are interpreted as linear mappings. We collect together some of its most useful properties in the next proposition. In particular, we see that if a linear mapping on a finite dimensional space is represented by a matrix, then the trace coincides with the usual definition of the trace of a square matrix. We recall that the transpose of a linear mapping $S \colon X \to Y$ is the linear mapping $S^t \colon Y^\sharp \to X^\sharp$ given by $\langle x, S^t\psi\rangle = \langle Sx, \psi\rangle$.

Proposition 1.6. *Let X and Y be vector spaces.*

(a) *Let $U \colon X \to X$ be a linear mapping of finite rank. The transpose, U^t, also has finite rank and $\mathrm{tr}_X\, U = \mathrm{tr}_{X^\sharp}\, U^t$.*

(b) *Let $S: X \to Y$ and $T: Y \to X$ be linear mappings, at least one of which is of finite rank. Then ST and TS have finite rank and*

$$\operatorname{tr}_Y ST = \operatorname{tr}_X TS\,.$$

(c) *Suppose X is finite dimensional and $V: X \to X$ is a linear mapping. If V is represented by the matrix A relative to some basis for X, then $\operatorname{tr}_X V$ is the sum of the diagonal entries of A.*

Proof. (a) If $\sum_{i=1}^n \varphi_i \otimes x_i$ is a representation of U, then $U^t = \sum_{i=1}^n x_i \otimes \varphi_i$ and so $\operatorname{tr}_{X^\sharp} U^t = \sum_{i=1}^n \langle x_i, \varphi_i \rangle = \operatorname{tr}_X U$.
(b) Suppose that $S = \sum_{i=1}^n \varphi_i \otimes y_i$ has finite rank. Then $TS = \sum_{i=1}^n \varphi_i \otimes Ty_i$ and $ST = \sum_{i=1}^n T^t\varphi_i \otimes y_i$ and so

$$\operatorname{tr}_X TS = \sum_{i=1}^n \langle Ty_i, \varphi_i \rangle = \sum_{i=1}^n \langle y_i, T^t\varphi_i \rangle = \operatorname{tr}_Y ST\,.$$

(c) Let $\{e_1, \ldots, e_n\}$ be a basis for X with coordinate functionals $\{\varphi_1, \ldots, \varphi_n\}$, so that each $x \in X$ has the expansion $x = \sum_{i=1}^n \varphi_i(x)e_i$. If $A = (a_{ij})$ is the matrix of V relative to this basis, then we have

$$V(x) = \sum_{j=1}^n \varphi_j(x)V(e_j) = \sum_{j=1}^n \sum_{i=1}^n \varphi_j(x)a_{ij}e_i\,.$$

Therefore $V = \sum_{i,j=1}^n \varphi_j \otimes a_{ij}e_i$ and so $\operatorname{tr}_X V = \sum_{i,j=1}^n a_{ij}\varphi_j(e_i) = \sum_{i=1}^n a_{ii}$. □

1.4 Tensor and Trace Duality

In this section we look again at the duality theory we have developed for tensor products. Recall that the dual space of the tensor product $X \otimes Y$ is the space $B(X \times Y)$ of bilinear forms on $X \times Y$ and that the tensor product $X^\sharp \otimes Y^\sharp$ is canonically embedded in the space $B(X \times Y)$. Now if X and Y are finite dimensional, then it is easy to see that $B(X \times Y) = X^\sharp \otimes Y^\sharp$, and so we have

$$(X \otimes Y)^\sharp = X^\sharp \otimes Y^\sharp\,,$$

with the duality relationship between the spaces $X \otimes Y$ and $X^\sharp \otimes Y^\sharp$ summarized by the relation

$$\langle x \otimes y, \varphi \otimes \psi \rangle = \varphi(x)\psi(y)\,.$$

In this case, each of the tensor products $X \otimes Y$ and $X^\sharp \otimes Y^\sharp$ is the dual of the other, and we have a very pleasing duality, which we shall refer to as *tensor duality*. It is interesting to look at this duality using spaces of linear mappings rather than tensor products. Continuing with our assumption of finite dimensionality, consider the space $L(X,Y)$ of linear mappings from X into Y. We may identify this space with the tensor product $X^\sharp \otimes Y$, and so its dual space is $X \otimes Y^\sharp$, which it is natural to interpret as $L(Y,X)$. Thus we have

$$L(X,Y)^\sharp = L(Y,X).$$

Let us examine the workings of the duality between these spaces. Suppose $S \in L(X,Y)$ and $T \in L(Y,X)$. Using the identifications described above, S and T correspond to tensors $u = \sum_{i=1}^n \varphi_i \otimes y_i \in X^\sharp \otimes Y$ and $v = \sum_{j=1}^m x_j \otimes \psi_j \in X \otimes Y^\sharp$ respectively. Then a straightforward computation shows that $\langle S,T\rangle = \langle u,v\rangle = \mathrm{tr}_Y\, ST$. Therefore the duality between the spaces $L(X,Y)$ and $L(Y,X)$ is given by

$$\langle S,T\rangle = \mathrm{tr}_Y\, ST = \mathrm{tr}_X\, TS.$$

We shall describe this relationship as *trace duality*.

We emphasize that tensor duality and trace duality are simply different ways of looking at the same thing, and we are free to work with whichever formalism best suits the context.

What can we say in infinite dimensions? If we drop the assumption that X and Y are finite dimensional, then in our description of tensor duality we must revert to our original identification of the dual space of $X \otimes Y$ with the space $B(X \times Y)$. This latter space will in general be strictly larger than $X^\sharp \otimes Y^\sharp$. And, in the language of trace duality, we must replace the space $L(X,Y)$ with the smaller space of linear mappings of finite rank, which we shall denote by $FL(X,Y)$. Thus in general we have

$$FL(X,Y)^\sharp = L(Y, X^{\sharp\sharp}),$$

with the duality given by $\langle S,T\rangle = \mathrm{tr}_Y\, ST = \mathrm{tr}_{X^{\sharp\sharp}}\, TS$, as in the finite dimensional case.

1.5 Examples and Applications

Spaces of Matrices

Consider the tensor product $\mathbb{K}^n \otimes \mathbb{K}^m$, where $\mathbb{K}^n$ and $\mathbb{K}^m$ are endowed with the standard bases, which we denote by $\{e_1,\dots,e_n\}$ and $\{f_1,\dots,f_m\}$ respectively. This tensor product can be identified with the vector space $M_{mn}(\mathbb{K})$ of $m \times n$ matrices with entries in the field $\mathbb{K}$. The tensor $e_i \otimes f_j$ is identified with the matrix that has the entry 1 in the (i,j) position and zeros elsewhere. Thus the tensor product basis $\{e_i \otimes f_j\}$ corresponds to the standard basis for $M_{mn}(\mathbb{K})$.

Vector Valued Functions

We now describe a process whereby a vector space of scalar valued functions can be converted into a space of vector valued functions. We begin with a very general situation and then look at some examples. Let $F(S)$ be the vector space of all functions from a set S into the scalar field $\mathbb{K}$, with the usual pointwise definitions of addition and multiplication by scalars. Now let X be a vector space and let $F(S,X)$ be the vector space of all functions from S into X. For each $f \in F(S)$ and $x \in X$ we may define a function from S into X by $s \mapsto f(s)x$. Let us for the moment write this function as $f \cdot x$. Consider the bilinear form $(f,x) \in F(S) \times X \mapsto f \cdot x \in F(S,X)$. Linearizing, we obtain a linear mapping

$$\sum_{i=1}^{n} f_i \otimes x_i \in F(S) \otimes X \mapsto \sum_{i=1}^{n} f_i \cdot x_i \in F(S,X).$$

Let us show that this mapping is injective. To this end, suppose that the function $\sum_{i=1}^{n} f_i \cdot x_i$ is zero. Then we have $\sum_{i=1}^{n} f_i(s)x_i = 0$ for every $s \in S$. But the evaluation functionals, $f \mapsto f(s)$, clearly form a separating subset of the dual space of $F(S)$ and so, by the remarks following Proposition 1.2, we have $\sum_{i=1}^{n} f_i \otimes x_i = 0$ in $F(S) \otimes X$.

Thus, we have an embedding of $F(S) \otimes X$ into $F(S,X)$ – taking the tensor product of the function space $F(S)$ with the space X has the effect of "adjoining" vector values to the functions in $F(S)$. In the light of this identification, we may dispense with the notation introduced above, so the tensor $\sum_{i=1}^{n} f_i \otimes x_i$ may also be taken to represent the function $s \mapsto \sum_{i=1}^{n} f_i(s)x_i$.

The same ideas can be applied to any subspace, $G(S)$, of $F(S)$ to produce a "vector valued version" of $G(S)$. For example, let $P(\mathbb{K})$ be the vector space of polynomial functions on $\mathbb{K}$ and let $P(\mathbb{K},X)$ be the space of polynomial functions on $\mathbb{K}$ with values in the vector space X. Each element P of $P(\mathbb{K},X)$ has the form $P(s) = \sum_{k=0}^{n} s^k x_k$ where $x_k \in X$. Let p_k be the monomial functions, $p_k(s) = s^k$. Then the polynomial P is given by the tensor $\sum_{k=0}^{n} p_k \otimes x_k$. On the other hand, it is easy to see that every element of the tensor product $P(\mathbb{K}) \otimes X$ yields a polynomial function from $\mathbb{K}$ into X. Thus we have a representation of the space $P(\mathbb{K},X)$ as a tensor product:

$$P(\mathbb{K},X) = P(\mathbb{K}) \otimes X.$$

For another example, let Ω be the vector space of all scalar sequences that are eventually zero. Thus, a typical element of Ω has the form $\omega = (s_1,\ldots,s_n,0,0,\ldots)$. Then the elements of the tensor product $\Omega \otimes X$ can be considered as sequences in X that are eventually zero. Let $\Omega(X)$ denote the space of all sequences in X that are eventually zero. If $x \in \Omega(X)$, then $x = (x_1,\ldots,x_n,0,0,\ldots) = \sum_{i=1}^{n} e_i \otimes x_i$, where e_i is the scalar sequence that has 1 in the n-th place and zero elsewhere. Thus we see that

$$\Omega(X) = \Omega \otimes X.$$

Vector Valued Measures

Let $\mathcal{A}$ be an algebra of subsets of a set S. We denote by $FM(\mathcal{A})$ the vector space of finitely additive scalar valued measures on $\mathcal{A}$. Thus, an element, μ, of $FM(\mathcal{A})$ is a function from $\mathcal{A}$ into the scalar field such that $\mu(\emptyset) = 0$ and $\mu(\cup_{i=1}^n A_i) = \sum_{i=1}^n \mu(A_i)$ for every finite, pairwise disjoint collection of sets $A_1, \ldots, A_n$ that belong to $\mathcal{A}$. For each vector space, X, we can define in just the same way the vector space $FM(\mathcal{A}, X)$ of finitely additive X-valued measures on $\mathcal{A}$. Proceeding just as above, we then have an embedding of the tensor product $FM(\mathcal{A}) \otimes X$ into $FM(\mathcal{A}, X)$ under which the tensor $\mu = \sum_{i=1}^n \mu_i \otimes x_i$ corresponds to the vector valued measure

$$\mu(A) = \sum_{i=1}^n \mu_i(A) x_i .$$

We leave it to the reader to verify that if X is finite dimensional, then $FM(\mathcal{A}) \otimes X = FM(\mathcal{A}, X)$.

Functions of Two Variables

We have already seen that the tensor product of dual spaces, $X^\sharp \otimes Y^\sharp$, can be embedded in the space $B(X \times Y)$ of bilinear forms. This process can be generalized, so that the tensor product of two function spaces can be interpreted as a space of functions of two variables. Thus, if S and T are two sets, then there is an injective linear mapping from $F(S) \otimes F(T)$ into $F(S \times T)$ which associates with the tensor $\sum_{i=1}^n f_i \otimes g_i$ the function $(s,t) \mapsto \sum_{i=1}^n f_i(s) g_i(t)$.

In some cases, a space of functions of two variables may be represented exactly as a tensor product of function spaces of one variable. For example, let $P(\mathbb{K} \times \mathbb{K})$ be the vector space of polynomial functions in two variables. Then we have

$$P(\mathbb{K} \times \mathbb{K}) = P(\mathbb{K}) \otimes P(\mathbb{K}) .$$

1.6 Exercises

Exercise 1.1. Show that the tensor product $X \otimes Y$ contains isomorphic copies of both X and Y.

Exercise 1.2. Let $X = Y = \mathbb{R}^2$. Show that $2e_1 \otimes e_1 - 2e_2 \otimes e_1 + e_1 \otimes e_2 - e_2 \otimes e_2$ is an elementary tensor in $X \otimes Y$.

Exercise 1.3. Show that $X \otimes (Y \otimes Z)$ is isomorphic to $(X \otimes Y) \otimes Z$.

Exercise 1.4. Show that the tensor product mapping $S \otimes T$ is injective (respectively, surjective) if and only if both S and T are injective (respectively, surjective).

Exercise 1.5. Let E and F be subspaces of X and Y respectively. Show that $(X/E) \otimes (Y/F)$ is not, in general, the same as $X \otimes Y/E \otimes F$.

Exercise 1.6. Show that there is a linear mapping $C\colon (X \otimes Y^\sharp) \otimes (Y \otimes Z) \to X \otimes Z$ such that $C((x \otimes \psi) \otimes (y \otimes z)) = \psi(y)(x \otimes z)$.

Exercise 1.7. Show that the rank of a tensor $u \in X \otimes Y$ is the same as the rank of the linear mapping from $X^\sharp$ into Y associated with u.

Exercise 1.8. The *symmetric tensor product* $X \otimes_s X$ is the subspace of $X \otimes X$ spanned by the tensors $x \otimes_s y = (x \otimes y + y \otimes x)/2$. Show that $X \otimes_s X$ is a universal space for symmetric bilinear mappings on $X \times X$ in the following sense: $B\colon X \times X \to Z$ is a symmetric bilinear mapping if and only if there exists a unique linear mapping $T\colon X \otimes_s X \to Z$ such that $B(x, y) = T(x \otimes_s y)$ for every $x, y \in X$. (To say that B is symmetric means that $B(x, y) = B(y, x)$ for all $x, y \in X$.)

Exercise 1.9. The *alternating tensor product* $X \otimes_a X$ is the subspace of $X \otimes X$ spanned by the tensors $x \otimes_a y = (x \otimes y - x \otimes y)/2$. Show that $X \otimes X$ is the direct sum of the subspaces $X \otimes_s X$ and $X \otimes_a X$.

Exercise 1.10. Show that if X and Y are infinite dimensional, then $B(X \times Y)$ is strictly larger than $X^\sharp \otimes Y^\sharp$.

Exercise 1.11. Find conditions on the set S or the vector space X that will ensure that $F(S, X) = F(S) \otimes X$.

Exercise 1.12. Let $I\colon P(\mathbb{K}) \to P(\mathbb{K})$ be the indefinite integral operator: $I(\sum_{k=0}^n a_k t^k) = \sum_{k=0}^n (k+1)^{-1} a_k t^{k+1}$. Describe the mapping $I \otimes I$ if $P(\mathbb{K}) \otimes P(\mathbb{K})$ is identified with the vector space of polynomials in two variables.

2. The Projective Tensor Product

In this chapter we investigate the simplest way to norm the tensor product of two Banach spaces. The projective tensor product linearizes bounded bilinear mappings just as the algebraic tensor product linearizes bilinear mappings. The projective tensor product derives its name from the fact that it behaves well with respect to quotient space constructions. The projective tensor product of ℓ_1 with X gives a representation of the space of absolutely summable sequences in X and projective tensor products with $L_1(\mu)$ lead to a study of the Bochner integral for Banach space valued functions. We also introduce the class of $\mathcal{L}_1$-spaces, whose finite dimensional structure is like that of ℓ_1. We study some techniques that make use of the Rademacher functions, including the Khinchine inequality. Finally, interpreting the elements of a projective tensor product as bilinear forms or operators leads to the introduction of the concept of nuclearity.

2.1 The Projective Norm

Let X and Y be Banach spaces. How should we norm the tensor product $X \otimes Y$? Consider first the elementary tensors. It is natural to require that

$$\|x \otimes y\| \leq \|x\|\|y\| . \tag{2.1}$$

Now let u be any element of $X \otimes Y$. If $\sum_{i=1}^{n} x_i \otimes y_i$ is a representation of u, then it follows from the triangle inequality that the norm must satisfy

$$\|u\| \leq \sum_{i=1}^{n} \|x_i\|\|y_i\| .$$

Since this holds for every representation of u, it follows that

$$\|u\| \leq \inf\left\{\sum_{i=1}^{n} \|x_i\|\|y_i\|\right\},$$

the infimum being taken over all representations of u. The right hand side of this inequality is the biggest possible candidate for a "natural" norm on $X \otimes Y$. This norm is known as the *projective norm* and is defined as follows:

$$\pi(u) = \inf\left\{\sum_{i=1}^{n} \|x_i\|\|y_i\| : u = \sum_{i=1}^{n} x_i \otimes y_i\right\}. \tag{2.2}$$

If it is necessary to specify the component spaces in the tensor product, we shall denote this norm by $\pi_{X,Y}(u)$, or $\pi(u; X \otimes Y)$.

Proposition 2.1. *Let X and Y be Banach spaces. Then π is a norm on $X \otimes Y$ and $\pi(x \otimes y) = \|x\|\|y\|$ for every $x \in X$, $y \in Y$.*

Proof. First, we show that $\pi(\lambda u) = |\lambda|\pi(u)$. This is obvious when λ is zero, so suppose that $\lambda \neq 0$. If $u = \sum_{i=1}^{n} x_i \otimes y_i$ is a representation of u then $\lambda u = \sum_{i=1}^{n} (\lambda x_i) \otimes y_i$, and so we have $\pi(\lambda u) \leq \sum_{i=1}^{n} \|\lambda x_i\|\|y_i\| = |\lambda| \sum_{i=1}^{n} \|x_i\|\|y_i\|$. Since this holds for every representation of u, it follows that $\pi(\lambda u) \leq |\lambda|\pi(u)$. In the same way, we have $\pi(u) = \pi(\lambda^{-1}\lambda u) \leq |\lambda|^{-1}\pi(\lambda u)$, giving $|\lambda|\pi(u) \leq \pi(\lambda u)$. Therefore $\pi(\lambda u) = |\lambda|\pi(u)$.

Now, to prove that π satisfies the triangle inequality, let $u, v \in X \otimes Y$ and let $\varepsilon > 0$. It follows from the definition that we may choose representations $u = \sum_{i=1}^{n} x_i \otimes y_i$ and $v = \sum_{j=1}^{m} w_j \otimes z_j$ such that $\sum_{i=1}^{n} \|x_i\|\|y_i\| \leq \pi(u) + \varepsilon/2$ and $\sum_{j=1}^{m} \|w_j\|\|z_j\| \leq \pi(v) + \varepsilon/2$. Then $\sum_{i=1}^{n} x_i \otimes y_i + \sum_{j=1}^{m} w_j \otimes z_j$ is a representation of $u + v$ and so

$$\pi(u + v) \leq \sum_{i=1}^{n} \|x_i\|\|y_i\| + \sum_{j=1}^{m} \|w_j\|\|z_j\| \leq \pi(u) + \pi(v) + \varepsilon .$$

Since this holds for every $\varepsilon > 0$, we have $\pi(u + v) \leq \pi(u) + \pi(v)$.

Now suppose that $\pi(u) = 0$. Then, for every $\varepsilon > 0$ there is a representation $\sum_{i=1}^{n} x_i \otimes y_i$ of u such that $\sum_{i=1}^{n} \|x_i\|\|y_i\| \leq \varepsilon$. Hence, for every $\varphi \in X^*$, $\psi \in Y^*$ we have $|\sum_{i=1}^{n} \varphi(x_i)\psi(y_i)| \leq \varepsilon\|\varphi\|\|\psi\|$. Since the value of the sum $\sum_{i=1}^{n} \varphi(x_i)\psi(y_i)$ is independent of the representation of u, it follows that $\sum_{i=1}^{n} \varphi(x_i)\psi(y_i) = 0$. But the dual spaces X^*, Y^* are separating subsets of the respective algebraic duals and so, by Proposition 1.2, it follows that $u = 0$.

Finally, we must show that $\pi(x \otimes y) = \|x\|\|y\|$. On the one hand, it is clear that $\pi(x \otimes y) \leq \|x\|\|y\|$. Choose $\varphi \in B_{X^*}$, $\psi \in B_{Y^*}$ such that $\varphi(x) = \|x\|$ and $\psi(y) = \|y\|$. Consider the bounded bilinear form, B, on $X \times Y$ given by $B(w, z) = \varphi(w)\psi(z)$. The linearization of B is a linear functional on $X \otimes Y$ and

$$\left|\tilde{B}\left(\sum_{i=1}^{n} x_i \otimes y_i\right)\right| \leq \sum_{i=1}^{n} |\tilde{B}(x_i \otimes y_i)| = \sum_{i=1}^{n} |\varphi(x_i)\psi(y_i)| \leq \sum_{i=1}^{n} \|x_i\|\|y_i\| ,$$

which implies that $|\tilde{B}(u)| \leq \pi(u)$ for every $u \in X \otimes Y$. Therefore $\tilde{B}$ is a bounded linear functional on the normed space $(X \otimes Y, \pi)$ of norm at most one. Hence $\|x\|\|y\| = \tilde{B}(x \otimes y) \leq \pi(x \otimes y)$ and we are done. □

We shall denote by $X \otimes_\pi Y$ the tensor product $X \otimes Y$ endowed with the projective norm, π. Unless the spaces X and Y are finite dimensional, this space is not complete. We denote its completion by $X\hat{\otimes}_\pi Y$. The Banach space $X\hat{\otimes}_\pi Y$ will be referred to as the *projective tensor product* of the Banach spaces X, Y.

We have seen in the proof of Proposition 1.2 that, in order to establish that a tensor u in $X \otimes_\pi Y$ is zero, it suffices to show that $\sum_{i=1}^n \varphi(x_i)\psi(y_i) = 0$ for every $\varphi \in X^*$, $\psi \in Y^*$, where $\sum_{i=1}^n x_i \otimes y_i$ is any representation of u. The question of identifying the zero tensor in the *completed* tensor product $X\hat{\otimes}_\pi Y$ is a more delicate matter, to which we will return in Chapter 4.

If A, B are subsets of X, Y respectively, let us write $A \otimes B$ for the set $\{x \otimes y : x \in A, y \in B\}$. This notation, though convenient, is rather dangerous, and will be used sparingly. We denote the convex hull of a set S by $\mathrm{co}(S)$ and its closure by $\overline{\mathrm{co}}(S)$.

Proposition 2.2. *The closed unit ball of $X\hat{\otimes}_\pi Y$ is the closed convex hull of the set $B_X \otimes B_Y$.*

Proof. The closed unit ball of $X\hat{\otimes}_\pi Y$ is the closure of the closed unit ball of the uncompleted tensor product $X \otimes_\pi Y$, and so it suffices to prove the proposition for the space $X \otimes_\pi Y$. Suppose that u is in the open unit ball of $X \otimes_\pi Y$. Then, by the definition of the projective norm, there is a representation of u of the form $\sum_{i=1}^n x_i \otimes y_i$, where x_i and y_i are all non-zero, and $\sum_{i=1}^n \|x_i\|\|y_i\| < 1$. Let $w_i = \|x_i\|^{-1}x_i$, $z_i = \|y_i\|^{-1}y_i$ and $\lambda_i = \|x_i\|\|y_i\|$. Then $u = \sum_{i=1}^n \lambda_i w_i \otimes z_i$, with $w_i \in B_X$, $z_i \in B_Y$, $\lambda_i \geq 0$ and $\sum_{i=1}^n \lambda_i < 1$. Therefore $u \in \mathrm{co}(B_X \otimes B_Y)$. It follows that the closed unit ball of $X \otimes Y$ is contained in $\overline{\mathrm{co}}(B_X \otimes B_Y)$. On the other hand, it is clear that $B_X \otimes B_Y$ is contained in the closed unit ball of $X \otimes_\pi Y$, and so the same is true for the closed convex hull of $B_X \otimes B_Y$. □

Next, we consider the tensor product of operators. Let $S \in \mathcal{L}(X, W)$ and $T \in \mathcal{L}(Y, Z)$. We have seen in Chapter 1 that there is a unique linear mapping $S \otimes T \colon X \otimes Y \to W \otimes Z$ such that $S \otimes T(x \otimes y) = (Sx) \otimes (Ty)$ for every $x \in X$, $y \in Y$. Let $u \in X \otimes Y$ and let $\sum_{i=1}^n x_i \otimes y_i$ be a representation of u. Then

$$\pi(S \otimes T(u)) = \pi\left(\sum_{i=1}^n (Sx_i) \otimes (Ty_i)\right) \leq \|S\|\|T\| \sum_{i=1}^n \|x_i\|\|y_i\|,$$

from which it follows that $\pi(S \otimes T(u)) \leq \|S\|\|T\|\pi(u)$. Therefore $S \otimes T$ is bounded for the projective norms on $X \otimes Y$ and $W \otimes Z$ and $\|S \otimes T\| \leq \|S\|\|T\|$. On the other hand, it follows easily from $S \otimes T(x \otimes y) = (Sx) \otimes (Ty)$ that $\|S \otimes T\| \geq \|S\|\|T\|$. Thus, we have $\|S \otimes T\| = \|S\|\|T\|$. The last step in this construction is to take the unique bounded extension of the operator $S \otimes T$ to the completions of $X \otimes_\pi Y$ and $W \otimes_\pi Z$. We shall denote this operator by $S \otimes_\pi T$. We summarize the properties of this process:

Proposition 2.3. *Let $S: X \to W$ and $T: Y \to Z$ be operators. Then there is a unique operator $S \otimes_\pi T: X \hat{\otimes}_\pi Y \to W \hat{\otimes}_\pi Z$ such that $S \otimes_\pi T(x \otimes y) = (Sx) \otimes (Ty)$ for every $x \in X$, $y \in Y$. Furthermore, $\|S \otimes_\pi T\| = \|S\|\|T\|$.*

The projective tensor product does not respect subspaces. In other words, if W is a subspace of the Banach space X, so that $W \otimes Y$ is an algebraic subspace of $X \otimes Y$, then the norm induced on $W \otimes Y$ by $X \otimes_\pi Y$ is not, in general, the projective norm $\pi_{W,Y}(.)$. To understand why this is so, consider the definition of the norm $\pi(u; W \otimes Y)$, where $u \in W \otimes Y$. The infimum that defines this norm is taken over the set of all representations of u in $W \otimes Y$. If we enlarge the space W to X, then the set of representations of u becomes bigger, and so we have

$$W \subset X \Rightarrow \pi(u; X \otimes Y) \leq \pi(u; W \otimes Y).$$

We shall explore this phenomenon later. For now, let us note a positive result for complemented subspaces.

Proposition 2.4. *Let E, F be complemented subspaces of X, Y respectively. Then $E \otimes F$ is complemented in $X \otimes_\pi Y$ and the norm on $E \otimes F$ induced by the projective norm of $X \otimes_\pi Y$ is equivalent to the projective norm $\pi_{E,F}$. If E and F are complemented by projections of norm one, then $E \otimes_\pi F$ is a subspace of $X \otimes_\pi Y$ and is also complemented by a projection of norm one.*

Proof. Let P, Q be projections from X, Y onto E, F respectively. It is easy to see that $P \otimes Q$ is a projection of $X \otimes_\pi Y$ onto $E \otimes F$. Let $u \in E \otimes F$. We have seen that $\pi_{X,Y}(u) \leq \pi_{E,F}(u)$. Now let $\sum_{i=1}^n x_i \otimes y_i$ be a representation of u in $X \otimes Y$. Then $u = P \otimes Q(u)$ and so $\sum_{i=1}^n (Px_i) \otimes (Qy_i)$ is a representation of u in $E \otimes F$. Therefore

$$\pi_{E,F}(u) \leq \sum_{i=1}^n \|Px_i\|\|Qy_i\| \leq \|P\|\|Q\| \sum_{i=1}^n \|x_i\|\|y_i\|.$$

Since this holds for every representation of u in $X \otimes Y$, it follows that

$$\pi_{X,Y}(u) \leq \pi_{E,F}(u) \leq \|P\|\|Q\| \pi_{X,Y}(u).$$

If E and F are complemented by projections of norm one, then we have $\pi_{E,F}(u) = \pi_{X,Y}(u)$ for every $u \in E \otimes F$. Also, $\|P \otimes Q\| = \|P\|\|Q\| = 1$, and hence $P \otimes Q$ is a projection of norm one onto $E \otimes F$. □

Taking completions, we see that these results also hold for the complete projective tensor products.

We now prove a result that explains the choice of the name "projective" for the norm π. We say that an operator $Q: Z \to Y$ is a *quotient operator* if Q is surjective and $\|y\| = \inf\{\|z\| : z \in Z, Q(z) = y\}$ for every $y \in Y$, or, equivalently, Q maps the open unit ball of Z onto the open unit ball of Y. This means simply that Y is isometrically isomorphic to the quotient space $Z/\ker Q$.

Proposition 2.5. *Let $Q\colon W \to X$ and $R\colon Z \to Y$ be quotient operators. Then $Q \otimes_\pi R\colon W\hat{\otimes}_\pi Z \to X\hat{\otimes}_\pi Y$ is a quotient operator.*

Proof. It suffices to show that $Q \otimes R\colon W \otimes_\pi Z \to X \otimes_\pi Y$ is a quotient operator. To see that $Q \otimes R$ is surjective, let $\sum_{i=1}^n x_i \otimes y_i \in X \otimes_\pi Y$. There exist $w_i \in W$ and $z_i \in Z$ such that $Q(w_i) = x_i$ and $R(z_i) = y_i$ for each i. Then $Q \otimes R(\sum_{i=1}^n w_i \otimes z_i) = \sum_{i=1}^n x_i \otimes y_i$. Therefore $Q \otimes R$ is surjective.

Now let $u \in X \otimes_\pi Y$. If $Q \otimes R(v) = u$, then $\pi(u) \le \|Q\|\|R\|\pi(v) = \pi(v)$. Given $\varepsilon > 0$, choose a representation $\sum_{i=1}^n x_i \otimes y_i$ of u such that $\sum_{i=1}^n \|x_i\|\|y_i\| \le \pi(u) + \varepsilon$. Now for each i, choose $w_i \in W$ and $z_i \in Z$ such that $Q(w_i) = x_i$, $R(z_i) = y_i$ and $\|w_i\| \le (1 + 2^{-n}\varepsilon)\|x_i\|$, $\|z_i\| \le (1 + 2^{-n}\varepsilon)\|y_i\|$. Then $Q \otimes R(\sum_{i=1}^n w_i \otimes z_i) = u$ and, using the fact that $\prod_{i=1}^n (1 + a_i) \le \exp(\sum_{i=1}^n |a_i|)$, we have

$$\pi\left(\sum_{i=1}^n w_i \otimes z_i\right) \le e^{4\varepsilon}\left(\sum_{i=1}^n \|x_i\|\|y_i\|\right) \le e^{4\varepsilon}(\pi(u) + \varepsilon).$$

Since this holds for every $\varepsilon > 0$, it follows that $\pi(u) = \inf\{\pi(v) : v \in W \otimes_\pi Z, Q \otimes R(v) = u\}$. □

In general, the projective norm, $\pi(u)$, of a tensor u can be quite difficult to compute. The definition of this norm requires that we examine every possible representation of u, and this is clearly not a practical proposition. In the next section we shall investigate the duality theory of projective tensor products and this will yield an alternative approach. However, there is one space whose tensor products are relatively easy to describe:

Example 2.6. *The projective tensor product $\ell_1\hat{\otimes}_\pi X$.*

Let X be a Banach space. We recall from Chapter 1 that the elements of the tensor product $\ell_1 \otimes X$ can be viewed as X-valued sequences. Under this identification, for each $a = (a_n) \in \ell_1$ and $x \in X$, the elementary tensor $a \otimes x$ corresponds to the sequence $(a_n x)$ in X. Now we observe that this sequence is absolutely summable, since $\sum_{n=1}^\infty \|a_n x\| \le (\sum_{n=1}^\infty |a_n|)\|x\|$. In other words, $a\otimes x$ belongs to the Banach space $\ell_1(X)$ of all absolutely summable sequences in X, where the norm is given by

$$\|(x_n)\|_1 = \sum_{n=1}^\infty \|x_n\|.$$

Thus there exists a linear mapping $J\colon \ell_1 \otimes X \to \ell_1(X)$ satisfying $J(a \otimes x) = (a_n x)$. Now, if $\sum_{i=1}^m a_i \otimes x_i$ is a representation of $u \in \ell_1 \otimes X$, where $a_i = (a_{in})_n$ for each i, then

$$\begin{aligned}\|J(u)\|_1 &= \left\|\left(\sum_{i=1}^m a_{in} x_i\right)_n\right\|_1 = \sum_{n=1}^\infty \left\|\left(\sum_{i=1}^m a_{in} x_i\right)_n\right\| \\ &\le \sum_{n=1}^\infty \sum_{i=1}^m \|a_{in} x_i\| = \sum_{i=1}^m \left(\sum_{n=1}^\infty |a_{in}|\right)\|x_i\| = \sum_{i=1}^m \|a_i\|\|x_i\|,\end{aligned}$$

and, since this holds for every representation of u, it follows that

$$\|J(u)\|_1 \leq \pi(u).$$

To prove the reverse inequality, fix a representation $\sum_{i=1}^m a_i \otimes x_i$ of u. Then $J(u) = (u_n)$, where $u_n = \sum_{i=1}^m a_{in} x_i$. We claim that the series $\sum_{n=1}^\infty e_n \otimes u_n$ converges to u in $\ell_1 \otimes_\pi X$, where $\{e_n\}$ is the standard unit vector basis for ℓ_1. Let Π_k denote the projection of ℓ_1 onto the first k coordinates, so that $\Pi_k(a) = \sum_{n=1}^k a_n e_n$. We have $\Pi_k(a) \to a$ as $k \to \infty$ and hence

$$\begin{aligned}\pi\Big(u - \sum_{n=1}^k e_n \otimes u_n\Big) &= \pi\Big(\sum_{i=1}^m a_i \otimes x_i - \sum_{n=1}^k \sum_{i=1}^m e_n \otimes a_{in} x_i\Big)\\ &= \pi\Big(\sum_{i=1}^m \Big(a_i \otimes x_i - \sum_{n=1}^k a_{in} e_n \otimes x_i\Big)\Big) = \pi\Big(\sum_{i=1}^m (a_i - \Pi_k a_i) \otimes x_i\Big)\\ &\leq \sum_{i=1}^m \|a_i - \Pi_k a_i\| \|x_i\|,\end{aligned}$$

from which it follows that $\pi(u - \sum_{n=1}^k e_n \otimes u_n) \to 0$ as $k \to \infty$. This proves our claim. We now have

$$\pi(u) = \pi\Big(\sum_{n=1}^\infty e_n \otimes u_n\Big) \leq \sum_{n=1}^\infty \|u_n\| = \|J(u)\|_1.$$

Therefore the linear mapping $J\colon \ell_1 \otimes_\pi X \to \ell_1(X)$ is an isometry. Now $\ell_1(X)$ is complete and so J extends to a unique isometric operator from the completed projective tensor product $\ell_1 \hat{\otimes}_\pi X$ into $\ell_1(X)$. Furthermore, this operator is surjective. To establish this, let $v = (x_n) \in \ell_1(X)$. We shall show that the series $\sum_{n=1}^\infty e_n \otimes x_n$ converges in $\ell_1 \hat{\otimes}_\pi X$. It is then easy to see that the sum of this series is mapped by J to u. Since we are working in the Banach space $\ell_1 \hat{\otimes}_\pi X$, it is enough to show that this series is Cauchy. But this follows immediately from $\pi(\sum_{n=j}^k e_n \otimes x_n) \leq \sum_{n=j}^k \|x_n\|$. To summarize, we have proved that there is a canonical isometric isomorphism

$$J\colon \ell_1 \hat{\otimes}_\pi X \to \ell_1(X),$$

which allows us to identify these spaces.

The arguments we have given for the space ℓ_1 apply equally well to the space $\ell_1(I)$, where I is an arbitrary indexing set, to give an identification

$$\ell_1(I) \hat{\otimes}_\pi X = \ell_1(I, X),$$

where $\ell_1(I, X)$ is the Banach space of absolutely summable families in X indexed by I, with the norm $\|(x_i)\|_1 = \sum_{i \in I} \|x_i\|$. To prove this, we need only recall that if (x_i) is an absolutely summable family, then there is a countable subset, I_0, of I such that $x_i = 0$ if $i \notin I_0$.

We can draw some interesting conclusions from the identification of $\ell_1(I)\hat{\otimes}_\pi X$ with $\ell_1(I, X)$. Recall that, in general, the projective tensor product does not respect subspaces – if W is a subspace of X, then $W\hat{\otimes}_\pi Y$ need not be a subspace of $X\hat{\otimes}_\pi Y$. The situation is different if one of the spaces is $\ell_1(I)$.

Proposition 2.7. *Let X be a Banach space and let Y be a closed subspace of X. Then $\ell_1(I)\hat{\otimes}_\pi Y$ is a subspace of $\ell_1(I)\hat{\otimes}_\pi X$.*

Proof. We have $\ell_1(I)\hat{\otimes}_\pi Y = \ell_1(I, Y)$ and $\ell_1(I)\hat{\otimes}_\pi X = \ell_1(I, X)$ and it is clear that $\ell_1(I, Y)$ is a subspace of $\ell_1(I, X)$. □

Our second application is a useful representation of the elements of a complete projective tensor product, $X\hat{\otimes}_\pi Y$, which complements nicely the representation of the elements of the algebraic tensor product $X \otimes Y$. We recall that every Banach space is a quotient of $\ell_1(I)$ for some suitably chosen indexing set I.

Proposition 2.8. *Let X and Y be Banach spaces. Let $u \in X\hat{\otimes}_\pi Y$ and $\varepsilon > 0$. Then there exist bounded sequences (x_n), (y_n) in X, Y respectively such that the series $\sum_{n=1}^\infty x_n \otimes y_n$ converges to u and*

$$\sum_{n=1}^{\infty} \|x_n\|\|y_n\| < \pi(u) + \varepsilon\,.$$

Proof. Choose an indexing set J and a quotient operator $Q\colon \ell_1(J) \to X$. Let I be the identity operator on Y. By Proposition 2.5, the tensor product operator $Q\otimes_\pi I\colon \ell_1(J)\hat{\otimes}_\pi Y \to X\hat{\otimes}_\pi Y$ is a quotient operator. Therefore there exists $v \in \ell_1(J)\hat{\otimes}_\pi Y$ such that $Q \otimes_\pi I(v) = u$ and $\pi(v) \leq \pi(u) + \varepsilon$. Since $\ell_1(J)\hat{\otimes}_\pi Y = \ell_1(J, Y)$, we may identify v with an absolutely summable family, (v_j) in Y. Now there exists a countable subset $J_0 = \{j_n\}$ of J such that $v_j = 0$ if $j \notin J_0$. Hence we may write v as $\sum_{n=1}^\infty e_{j_n} \otimes v_{j_n}$, with $\pi(v) = \sum_{n=1}^\infty \|v_{j_n}\|$. Let $x_n = Q(e_{j_n})$ and $y_n = v_{j_n}$. Then we have $u = Q\otimes_\pi I(v) = \sum_{n=1}^\infty x_n \otimes y_n$ and $\sum_{n=1}^\infty \|x_n\|\|y_n\| = \pi(v) \leq \pi(u) + \varepsilon$. □

It follows immediately from this result that we have

$$\pi(u) = \inf\left\{\sum_{n=1}^{\infty} \|x_n\|\|y_n\| : \sum_{n=1}^{\infty} \|x_n\|\|y_n\| < \infty,\ u = \sum_{n=1}^{\infty} x_n \otimes y_n\right\},$$

the infimum being taken over all the representations of $u \in X\hat{\otimes}_\pi Y$ as described in the statement of the proposition. There are some minor variations of this formula that are of use. For example, we may write u in the form $\sum_{n=1}^\infty \lambda_n x_n \otimes y_n$ where $\|x_n\| = \|y_n\| = 1$ for every n, or even as $\sum_{n=1}^\infty \lambda_n x_n \otimes y_n$ where $x_n \to 0$ and $y_n \to 0$. We have corresponding formulas for the projective norm:

$$\pi(u) = \inf\left\{\sum_{n=1}^{\infty}|\lambda_n| : u = \sum_{n=1}^{\infty}\lambda_n x_n \otimes y_n,\ \sum_{n=1}^{\infty}|\lambda_n| < \infty,\ \|x_n\| = \|y_n\| = 1\right\}$$
$$= \inf\left\{\sum_{n=1}^{\infty}|\lambda_n|\|x_n\|\|y_n\| : u = \sum_{n=1}^{\infty}\lambda_n x_n \otimes y_n,\ \sum_{n=1}^{\infty}|\lambda_n| < \infty,\ x_n, y_n \to 0\right\}.$$

2.2 The Dual Space of $X\hat{\otimes}_\pi Y$.

We have seen in Chapter 1 that the tensor product $X \otimes Y$ linearizes bilinear mappings on $X \times Y$. We now add norms to this picture. Recall that a bilinear mapping $B\colon X \times Y \to Z$ is said to be *bounded* if there exists a positive constant, C, such that $\|B(x,y)\| \le C\|x\|\|y\|$ for every $x \in X$, $y \in Y$, or equivalently, if B is bounded on the product $B_X \times B_Y$ of the unit balls. We denote by $\mathcal{B}(X \times Y, Z)$ the Banach space of bounded bilinear mappings from $X \times Y$ into Z, where the norm is given by $\|B\| = \sup\{\|B(x,y)\| : x \in B_X, y \in B_Y\}$. When Z is the field of scalars, we denote this space by $\mathcal{B}(X \times Y)$.

Theorem 2.9. *Let $B\colon X \times Y \to Z$ be a bounded bilinear mapping. Then there exists a unique operator $\tilde{B}\colon X\hat{\otimes}_\pi Y \to Z$ satisfying $\tilde{B}(x \otimes y) = B(x,y)$ for every $x \in X$, $y \in Y$. The correspondence $B \longleftrightarrow \tilde{B}$ is an isometric isomorphism between the Banach spaces $\mathcal{B}(X \times Y, Z)$ and $\mathcal{L}(X\hat{\otimes}_\pi Y, Z)$.*

Proof. There exists a unique linear mapping $\tilde{B}\colon X \otimes Y \to Z$ such that $\tilde{B}(x \otimes y) = B(x,y)$ for every x,y. Let us show first that $\tilde{B}$ is bounded for the projective norm on $X \otimes Y$. For $u = \sum_{i=1}^{n} x_i \otimes y_i \in X \otimes Y$ we have

$$\|\tilde{B}(u)\| = \left\|\sum_{i=1}^{n} B(x_i, y_i)\right\| \le \|B\|\sum_{i=1}^{n}\|x_i\|\|y_i\|.$$

Since this holds for every representation of u, it follows that $\|\tilde{B}(u)\| \le \|B\|\pi(u)$. Therefore $\tilde{B}$ is bounded and satisfies $\|\tilde{B}\| \le \|B\|$. On the other hand, $\|B(x,y)\| = \|\tilde{B}(x \otimes y)\| \le \|\tilde{B}\|\|x\|\|y\|$ gives $\|B\| \le \|\tilde{B}\|$. Therefore $\|B\| = \|\tilde{B}\|$. Now the operator $\tilde{B}\colon X \otimes_\pi Y \to Z$ has a unique extension to an operator $\tilde{B}\colon X\hat{\otimes}_\pi Y \to Z$ with the same norm. The mapping $B \mapsto \tilde{B}$ is clearly a linear isometry. It only remains to show that this mapping is surjective. Let $L \in \mathcal{L}(X\hat{\otimes}_\pi Y, Z)$. The bounded bilinear mapping defined by $B(x,y) = L(x \otimes y)$ satisfies $\tilde{B} = L$. □

Thus we have a canonical identification

$$\mathcal{B}(X \times Y, Z) = \mathcal{L}(X\hat{\otimes}_\pi Y, Z).$$

If we take Z to be the scalar field, we obtain a canonical identification of the dual space of the projective tensor product with the space of bounded bilinear forms:

$$\mathcal{B}(X \times Y) = (X\hat{\otimes}_\pi Y)^* .$$

With this identification, the action of a bounded bilinear form B as a bounded linear functional on $X\hat{\otimes}_\pi Y$ is given by

$$\left\langle \sum_{i=1}^{n} x_i \otimes y_i, B \right\rangle = \sum_{i=1}^{n} B(x_i, y_i) .$$

This duality yields a new formula for the projective norm:

$$\pi(u) = \sup\{|\langle u, B\rangle| : B \in \mathcal{B}(X \times Y), \|B\| \leq 1\} . \tag{2.3}$$

We now have two ways to compute a projective norm; the original definition and the duality formula above. The next example shows how effective this approach can be.

Example 2.10. *The tensor diagonal in $\ell_2\hat{\otimes}_\pi\ell_2$.*

It is not an easy matter to give a simple description of the space $\ell_2\hat{\otimes}_\pi\ell_2$. However, this space contains a familiar subspace whose presence is rather surprising. By the *tensor diagonal* in $\ell_2\hat{\otimes}_\pi\ell_2$ we mean the closed subspace, D, that is generated by the tensors $e_n \otimes e_n$, where $\{e_n\}$ is the standard unit vector basis for ℓ_2. Let us compute the projective norm of an element u of D of the form $u = \sum_{n=1}^k \alpha_n e_n \otimes e_n$. By the definition of this norm, we have $\pi(u) \leq \sum_{n=1}^k |\alpha_n|$. Now consider the bilinear form on $\ell_2 \times \ell_2$ defined by

$$B(x, y) = \sum_{n=1}^{k} \operatorname{sgn}(\alpha_n) x_n y_n ,$$

where $\operatorname{sgn}(\alpha)$ is a scalar of absolute value one such that $\operatorname{sgn}(\alpha)\alpha = |\alpha|$. It is easy to see that $\|B\| = 1$. Therefore, by the duality formula (2.3),

$$\pi(u) \geq \langle u, B\rangle = \sum_{n=1}^{k} \alpha_n B(e_n, e_n) = \sum_{n=1}^{k} |\alpha_n| .$$

Thus we have $\pi(u) = \sum_{n=1}^k |\alpha_n|$. It follows that D is isometrically isomorphic to ℓ_1.

This subspace is even complemented in $\ell_2\hat{\otimes}_\pi\ell_2$ by a projection of norm one. The projection is the linearization of the bilinear mapping from $\ell_2 \times \ell_2$ into D given by $(x, y) \mapsto \sum_{n=1}^\infty x_n y_n e_n \otimes e_n$. We leave it as an easy exercise to the reader that this operator has the stated properties.

The fact that $\ell_2\hat{\otimes}_\pi\ell_2$ contains a complemented isometric copy of ℓ_1 alerts us to some negative aspects of the projective tensor product. For example, although the space ℓ_2 is reflexive, $\ell_2\hat{\otimes}_\pi\ell_2$ is not.

The duality theory that we have developed between projective tensor products and spaces of bounded bilinear forms can also be formulated in terms of spaces of operators. With each bounded bilinear form $B \in \mathcal{B}(X \times Y)$ there is an associated operator $L_B \in \mathcal{L}(X, Y^*)$, defined by $\langle y, L_B(x) \rangle = B(x, y)$. It is straightforward to verify that the mapping $B \mapsto L_B$ is an isometric isomorphism between the spaces $\mathcal{B}(X \times Y)$ and $\mathcal{L}(X, Y^*)$. Thus, we have an identification:

$$(X \hat{\otimes}_\pi Y)^* = \mathcal{L}(X, Y^*),$$

through which the action of an operator $S \colon X \to Y^*$ as a linear functional on $X \hat{\otimes}_\pi Y$ is given by

$$\Big\langle \sum_{i=1}^n x_i \otimes y_i, S \Big\rangle = \sum_{i=1}^n \langle y_i, S x_i \rangle .$$

In the same way, we have another identification:

$$(X \hat{\otimes}_\pi Y)^* = \mathcal{L}(Y, X^*),$$

where the action of $T \in \mathcal{L}(Y, X^*)$ on $X \hat{\otimes}_\pi Y$ is given by

$$\Big\langle \sum_{i=1}^n x_i \otimes y_i, T \Big\rangle = \sum_{i=1}^n \langle x_i, T y_i \rangle .$$

These identifications yield two variations on the duality formula for the projective norm of an element of $X \hat{\otimes}_\pi Y$:

$$\begin{aligned} \pi(u) &= \sup\{|\langle u, S \rangle| : S \in \mathcal{L}(X, Y^*),\ \|S\| \le 1\} \\ &= \sup\{|\langle u, T \rangle| : T \in \mathcal{L}(Y, X^*),\ \|T\| \le 1\} . \end{aligned}$$

Of course, these are nothing more than restatements of (2.3). However, it is sometimes more convenient to work with operators than with bilinear forms.

With the assistance of this duality theory, we now look again at the question of when the projective tensor product respects subspaces. The following proposition is a straightforward consequence of the duality formula (2.3) and the Hahn–Banach Theorem.

Proposition 2.11. *Let W and Z be subspaces of X and Y respectively. Then $W \hat{\otimes}_\pi Z$ is a subspace of $X \hat{\otimes}_\pi Y$ if and only if every bounded bilinear form on $W \times Z$ extends to a bounded bilinear form on $X \times Y$ with the same norm.*

If we fix the second component space and use the operator version of duality, the existence of Hahn–Banach type extensions of bilinear forms translates into an extension property of operators:

Corollary 2.12. *Let W be a subspace of X. Then $W \hat{\otimes}_\pi Y$ is a subspace of $X \hat{\otimes}_\pi Y$ if and only if every operator from W into Y^* extends to an operator of the same norm from X into Y^*.*

A Banach space Z is said to be *injective* if it has the property that, for every Banach space X and every subspace W of X, every operator from W into Z extends to an operator from X into Z of the same norm. Equivalently, by considering the identity operator on Z, we see that Z is injective if and only if Z is complemented by a norm one projection in any Banach space that contains it as a subspace. It is not difficult to show that the space ℓ_∞ is injective. This can also be seen from the fact that projective tensor products with ℓ_1 respect subspaces. The corollary above shows that projective tensor products with the Banach space Y respect subspaces if and only if Y^* is an injective space.

There is one important setting in which subspaces and projective tensor products always work well together:

Theorem 2.13. *Every bounded bilinear form on $X \times Y$ has an extension to a bounded bilinear form on $X^{**} \times Y^{**}$ with the same norm.*

Proof. Let A be a bounded bilinear form on $X \times Y$ and let S be the associated operator from X into Y^*, so that $A(x, y) = \langle y, Sx\rangle$ for every $x \in X$, $y \in Y$. Consider the bounded bilinear form B on $X^{**} \times Y^{**}$ given by $B(x^{**}, y^{**}) = \langle S^*y^{**}, x^{**}\rangle$, where $S^*\colon Y^{**} \to X^*$ is the adjoint of S. If $x \in X$ and $y \in Y$, then

$$B(x, y) = \langle S^*y, x\rangle = \langle y, Sx\rangle = A(x, y),$$

and so B is an extension of A. Furthermore, $\|A\| = \|S\| = \|S^*\| = \|B\|$. □

Corollary 2.14. *$X\hat{\otimes}_\pi Y$ is a subspace of $X^{**}\hat{\otimes}_\pi Y^{**}$.*

2.3 $L_1(\mu)\hat{\otimes}_\pi X$ and the Bochner Integral

In this section, we shall identify the projective tensor product $L_1(\mu)\hat{\otimes}_\pi X$ with a Banach space of integrable vector valued functions. In order to do this, it is necessary first to put in place a suitable integration theory for functions with values in a Banach space. Throughout this section, (Ω, Σ, μ) will denote a measure space, by which we mean that Σ is a σ-algebra of subsets of the set Ω and μ is a positive measure on Σ, which we shall assume to be complete (if $E \in \Sigma$ has measure zero, then every subset of E belongs to Σ). We shall refer to the elements of Σ as measurable subsets of Ω.

We begin with the definition of a measurable function. A function $f\colon \Omega \to X$ is said to be a *simple function* if f has only a finite number of distinct values. Thus, there is a partition of Ω into disjoint subsets $E_1, \ldots, E_n$ and distinct nonzero elements $x_1, \ldots, x_n$ of X such that $f(\omega) = x_i$ if $\omega \in E_i$. We may then write f as $\sum_{i=1}^n \chi_{E_i} x_i$, where χ_E denotes the characteristic function of the set E. We call this the canonical representation of f. We say that f is a *μ-measurable simple function* if f is simple and each of the sets E_i is measurable. A function $f\colon \Omega \to X$ is said to be a *μ-measurable function* if there is a sequence (f_n) of μ-measurable simple functions that converges almost everywhere to f. We note that if f is μ-measurable, then the scalar function $\|f\|$ is also μ-measurable, since if (f_n) converges μ-almost everywhere to f, then $(\|f_n\|)$ converges μ-almost everywhere to $\|f\|$.

There are other, equally natural, approaches to the definition of measurability. A function $f\colon \Omega \to X$ is *weakly μ-measurable* if the scalar valued function φf is μ-measurable for every $\varphi \in X^*$. And we shall say that f is *Borel μ-measurable* if $f^{-1}(O)$ is a measurable set for every open subset O of X. The key ingredient that is needed to connect these concepts to the definition of μ-measurability is separability. The function $f\colon \Omega \to X$ will be said to be *μ-essentially separably valued* if there exists a measurable subset E of Ω whose complement has μ-measure zero, such that $f(E)$ is contained in a separable subspace Y of X. We also say that $f(\Omega)$ is μ-essentially contained in Y. We leave it as an exercise to the reader to show that if the measure μ is σ-finite, then $f\colon \Omega \to X$ is μ-measurable if and only if the function $\chi_E f$ is μ-measurable for every $E \in \Sigma$ of finite μ-measure.

Proposition 2.15 (Pettis Measurability Theorem). *Let (Ω, Σ, μ) be a σ-finite measure space. The following are equivalent for a function $f\colon \Omega \to X$:*

(i) *f is μ-measurable,*
(ii) *f is weakly μ-measurable and μ-essentially separably valued,*
(iii) *f is Borel μ-measurable and μ-essentially separably valued.*

Proof. (i) implies (ii): suppose that f is μ-measurable and let (f_n) be a sequence of μ-measurable simple functions that converges μ-almost everywhere to f. Then, for every $\varphi \in X^*$, (φf_n) is a sequence of scalar valued μ-measurable simple functions that converges μ-almost everywhere to φf, and hence φf is μ-measurable. Let Y be the closed subspace of X generated by the ranges of the functions f_n. Then Y is separable and $f(\Omega)$ is μ-essentially contained in Y.

(ii) implies (iii): suppose that f is a weakly μ-measurable function and that $f(\Omega)$ is μ-essentially contained in a separable subspace Y of X. We may assume without loss of generality that $f(\Omega)$ is contained in Y, since we may alter the values of f on a μ-null set without affecting either (ii) or (iii). Since the embedding of Y into X is continuous, it suffices to prove that f is Borel μ-measurable as a function from Ω into Y. The fact that φf is a μ-measurable function for every $\varphi \in X^*$ implies that f is Borel μ-measurable for the weak topology on Y. In other words, $f^{-1}(U)$ belongs to Σ for every weakly open subset U of Y. Hence the same is true for weakly closed subsets of Y. Now every closed ball in Y, being convex, is also weakly closed. And, by the separability of Y, every open subset U of Y is the union of a countable family of closed balls. It follows that $f^{-1}(U)$ belongs to Σ. Therefore f is Borel μ-measurable.

(iii) implies (i): suppose that f is Borel μ-measurable and that $f(\Omega)$ is essentially contained in a separable subspace Y of X. We may assume without loss of generality that $f(\Omega) \subset Y$. By the remarks preceding the statement, it suffices to consider the case where μ is finite. Let $\{y_k\}$ be a countable dense subset of Y. Then, for each n, the union of the open balls $B(y_k, 1/n)$ is Y. By the Borel measurability of f, the inverse images $E_k^n = f^{-1}(B(y_k, 1/n))$ are measurable and cover Ω for every n. We convert this into a disjoint measurable cover by setting $F_1^n = E_1^n$ and $F_k^n = E_k^n \setminus (E_1^n \cup \cdots \cup E_{k-1}^n)$ for $k > 1$. For each n, let $g_n = \sum_{k=1}^\infty \chi_{F_k^n} y_k$. Then we have $\|f(\omega) - g_n(\omega)\| < 1/n$ for every $\omega \in \Omega$. Therefore the sequence (g_n) of countably valued functions converges uniformly to f on Ω.

We truncate the expressions defining the functions g_n to get a sequence of μ-measurable simple functions that converges μ-almost everywhere to f as follows. For each n, choose m_n so that

$$\mu\Bigl(\bigcup_{k=m_n+1}^{\infty} F_k^n\Bigr) \le 2^{-n},$$

and let $h_n = \sum_{k=1}^{m_n} \chi_{F_k^n} y_k$, so that $\|f(\omega) - h_n(\omega)\| < 1/n$ for $\omega \in \Omega \setminus C_n$, where $C_n = \bigcup_{k=m_n+1}^\infty F_k^n$. Let $C = \bigcap_{n=1}^\infty \bigcup_{k=n}^\infty C_k$. Then C is a μ-null set and, if $\omega \notin C$, then there exists n such that $\omega \notin C_k$ for any $k \ge n$. Therefore $\|f(\omega) - h_n(\omega\| < 1/k \le 1/n$ for every $k \ge n$. □

This result can be applied to an arbitrary measure, provided the function f is supported by a set of σ-finite measure. Also, we can extend to the vector valued situation many of the standard properties of measurable scalar functions. For example, on a σ-finite measure space, the pointwise almost everywhere limit of a sequence of measurable functions is measurable. Egoroff's Theorem also holds for vector valued functions on a finite measure space; the proof is exactly as for the scalar case.

We can now define our integral. Let f be a μ-measurable simple function, with canonical representation $\sum_{i=1}^n \chi_{A_i} x_i$ and suppose that each of the sets A_i has finite μ-measure. The integral of f over a measurable subset E of Ω is defined to be

$$\int_E f\, d\mu = \sum_{i=1}^n \mu(A_i \cap E) x_i .$$

Now a μ-measurable function $f\colon \Omega \to X$ is said to be *Bochner integrable* if there exists a sequence (f_n) of μ-measurable simple functions converging μ almost everywhere to f and satisfying

$$\lim_{n\to\infty} \int_\Omega \|f - f_n\|\, d\mu = 0 .$$

The *Bochner integral* of f over a measurable subset E of Ω is

$$\int_E f\,d\mu = \lim_{n\to\infty}\int_E f_n\,d\mu\,.$$

The fact that this limit exists and is independent of the particular sequence (f_n) is proved exactly as in the case of scalar functions.

There is a very simple test for Bochner integrability:

Proposition 2.16 (Bochner's Theorem). *If $f\colon \Omega \to X$ is a μ-measurable function, then f is Bochner integrable if and only if the scalar function $\|f\|$ is integrable.*

Proof. Suppose that f is Bochner integrable. Let (f_n) be a sequence of μ-measurable simple functions converging μ-almost everywhere to f as in the above definition. Then

$$\int_\Omega \|f\|\,d\mu = \int_\Omega \|(f-f_n)+f_n\|\,d\mu \le \int_\Omega \|f-f_n\|\,d\mu + \int_\Omega \|f_n\|\,d\mu$$

for each n, and it follows that $\int_\Omega \|f\|\,d\mu$ is finite.

Conversely, suppose that $\|f\|$ is integrable. Let (f_n) be a sequence of μ-measurable simple functions that converges μ-almost everywhere to f. Fix $\delta > 0$ and let

$$g_n(\omega) = \begin{cases} f_n(\omega) & \text{if } \|f_n(\omega)\| \le (1+\delta)\|f(\omega)\|\,, \\ 0 & \text{otherwise}\,. \end{cases}$$

Now (g_n) is a sequence of μ-measurable simple functions that converges μ-almost everywhere to f. Furthermore, the sequence of scalar functions $(\|f-g_n\|)$ is dominated by the integrable function $(2+\delta)\|f\|$. Therefore, by the Dominated Convergence Theorem, we have $\int_\Omega \|f-g_n\|\,d\mu \to 0$. □

The Dominated Convergence Theorem is also valid for the Bochner integral. The proof is a straightforward application of Bochner's Theorem.

Corollary 2.17. *If $f\colon \Omega \to X$ is Bochner integrable then*

$$\left\|\int_\Omega f\,d\mu\right\| \le \int_\Omega \|f\|\,d\mu\,.$$

Proof. Let (f_n) be a sequence of μ-measurable simple functions converging μ-almost everywhere to f, with $\lim_n \int_\Omega f_n\,d\mu \to \int_\Omega f\,d\mu$. Then the sequence $(\|f_n\|)$ converges in μ-mean to $\|f\|$. Therefore

$$\begin{aligned} \left\|\int_\Omega f\,d\mu\right\| &= \left\|\lim_{n\to\infty}\int_\Omega f_n\,d\mu\right\| = \lim_{n\to\infty}\left\|\int_\Omega f_n\,d\mu\right\| \\ &\le \lim_{n\to\infty}\int_\Omega \|f_n\|\,d\mu = \int_\Omega \|f\|\,d\mu\,. \end{aligned}$$

□

Next, we show that operators respect the Bochner integral.

Proposition 2.18. *Let $f\colon \Omega \to X$ be Bochner integrable. If $T \in \mathcal{L}(X,Y)$ then the function $Tf\colon \Omega \to Y$ is Bochner integrable and*

$$\int_\Omega Tf\,d\mu = T\Big(\int_\Omega f\,d\mu\Big).$$

Proof. Let (f_n) be a sequence of μ-measurable simple functions such that (f_n) converges μ-almost everywhere to f and $\int_\Omega f_n\,d\mu \to \int_\Omega f\,d\mu$. It is easy to see that (Tf_n) is a sequence of μ-measurable simple functions that converges μ-almost everywhere to Tf and that $\int_\Omega Tf_n\,d\mu = T(\int_\Omega f_n\,d\mu)$ for every n. Furthermore, we have $\int_\Omega \|Tf - Tf_n\|\,d\mu \le \|T\| \int_\Omega \|f - f_n\|\,d\mu$. Therefore

$$\begin{aligned}\int_\Omega Tf\,d\mu &= \lim_{n\to\infty} \int_\Omega Tf_n\,d\mu = \lim_{n\to\infty} T\Big(\int_\Omega f_n\,d\mu\Big)\\ &= T\Big(\lim_{n\to\infty} \int_\Omega f_n\,d\mu\Big) = T\Big(\int_\Omega f\,d\mu\Big).\end{aligned}$$

□

The *Lebesgue–Bochner space* $L_1(\mu, X)$ is the Banach space of equivalence classes of Bochner integrable functions $f\colon \Omega \to X$, with the norm

$$\|f\|_1 = \int_\Omega \|f\|\,d\mu.$$

We follow the usual convention in using the symbol f for the equivalence class containing f. The completeness of the space $L_1(\mu, X)$ is proved exactly as in the case of the space $L_1(\mu)$.

Example 2.19. *The projective tensor product $L_1(\mu)\hat{\otimes}_\pi X$*

We shall now see that the projective tensor product $L_1(\mu)\hat{\otimes}_\pi X$ can be identified with the space $L_1(\mu, X)$. We begin with the bounded bilinear mapping of norm one from $L_1(\mu) \times X$ into $L_1(\mu, X)$ that maps (f, x) to the function $f \otimes x$. Linearizing, we obtain an operator

$$J\colon L_1(\mu)\hat{\otimes}_\pi X \to L_1(\mu, X)$$

of norm one. Thus, we have $\|Ju\|_1 \le \pi(u)$ for every $u \in L_1(\mu)\hat{\otimes}_\pi X$. On the other hand, let $f \in L_1(\mu, X)$ be a Bochner integrable simple function with canonical representation $\sum_{i=1}^n \chi_{A_i} x_i$. Then $f = Ju$ where $u = \sum_{i=1}^n \chi_{A_i} \otimes x_i$ and

$$\pi(u) \le \sum_{i=1}^n \|\chi_{A_i}\|\|x_i\| = \sum_{i=1}^n \mu(A_i)\|x_i\| = \|Ju\|_1.$$

Therefore $\pi(u) = \|Ju\|_1$ for such functions. Now let S be the normed subspace of $L_1(\mu)$ consisting of the integrable simple functions. Since S is dense in

$L_1(\mu)$, it follows that $S\otimes X$ is a dense subspace of $L_1(\mu)\hat{\otimes}_\pi X$ The argument we have given above shows that $\pi(u) = \|Ju\|_1$ for every $u \in S\otimes X$. Therefore J is an isometry. It only remains to show that J is surjective. But this follows immediately from the definition of the Bochner integral, since we have seen that the Bochner integrable simple functions are all in the range of J.

Taking μ to be the counting measure on $\mathbb{N}$, we obtain as a special case the identification $\ell_1\hat{\otimes}_\pi X = \ell_1(X)$ established previously. Like ℓ_1, projective tensor products with $L_1(\mu)$ respect subspaces: if Y is a subspace of X, then $L_1(\mu)\hat{\otimes}_\pi Y$ is a subspace of $L_1(\mu)\hat{\otimes}_\pi X$. This follows immediately from the fact that the L_1 norm of a Bochner integrable function with values in Y is unchanged if the range space is enlarged to X. Now if the measure μ is σ-finite, then the dual space of $L_1(\mu)$ is $L_\infty(\mu)$. We can deduce from the results of Section 2 that $L_\infty(\mu)$ is an injective space. In fact, this is true for every measure μ, but we shall not prove it here.

2.4 $\mathcal{L}_1$-spaces

We have observed that projective tensor products do not, in general, respect subspaces. This might suggest that it is not possible to deduce useful information about tensor products involving the Banach space X from a knowledge of tensor products with subspaces of X. In fact, even the finite dimensional subspaces of X carry vital information. To see this, let us fix Banach spaces X and Y, and an element u of the tensor product $X\otimes Y$. If M and N are subspaces of X and Y respectively such that u belongs to $M\otimes N$, then $\pi(u; X\otimes Y) \le \pi(u; M\otimes N)$. Now, given $\varepsilon > 0$, there exists a representation $\sum_{i=1}^n x_i\otimes y_i$ of u such that $\sum_{i=1}^n \|x_i\|\|y_i\| \le \pi(u; X\otimes Y)+\varepsilon$. Let M and N be the finite dimensional subspaces of X and Y spanned by $\{x_1,\ldots,x_n\}$ and $\{y_1,\ldots,y_n\}$ respectively. Then u lies in $M\otimes N$ and $\pi(u; M\otimes N) \le \sum_{i=1}^n \|x_i\|\|y_i\| \le \pi(u; X\otimes Y)+\varepsilon$. Therefore

$$\pi(u; X\otimes Y) = \inf\{\pi(u; M\otimes N) : u\in M\otimes N,\ \dim M,\ \dim N < \infty\}. \tag{2.4}$$

The infimum is taken over all the pairs of finite dimensional subspaces M, N, for which u belongs to $M\otimes N$. Thus the values of the projective norm are completely determined by the behaviour of the norm on finite dimensional spaces. The property described by the above equation is expressed by saying that the projective norm is *finitely generated.*

We now show one way in which this property can be exploited. We have already seen how well behaved the space $L_1(\mu)$ is in the formation of projective tensor products. Any space whose finite dimensional structure is like that of $L_1(\mu)$ will inherit some of that good behaviour. We now proceed to a description of such spaces. Let $\lambda > 1$. A Banach space X is said to be an $\mathcal{L}_{1,\lambda}$*-space* if every finite dimensional subspace M of X is contained in a finite dimensional subspace N, of dimension k, say, whose Banach–Mazur distance from the space ℓ_1^k is at most λ. This means that there exists an isomorphism $T\colon N \to \ell_1^k$ such that $\|T\|\|T^{-1}\| \le \lambda$. If X is a $\mathcal{L}_{1,\lambda}$-space for some $\lambda > 1$ then X is said to be an *$\mathcal{L}_1$-space.* We shall see shortly that every $L_1(\mu)$ is a $\mathcal{L}_1$-space, but first we show that this is a useful concept.

In order to state our main result about $\mathcal{L}_1$-spaces let us introduce some terminology. We shall say that *projective tensor products with X respect subspaces isomorphically* if, for every Banach space Y, and every subspace Z of Y, the projective norm on $X \otimes Z$ is equivalent to the norm induced by the projective norm on $X \otimes Y$.

Theorem 2.20. *If X is a $\mathcal{L}_1$-space then projective tensor products with X respect subspaces isomorphically.*

Proof. Suppose that X is a $\mathcal{L}_{1,\lambda}$-space. Let Z be a subspace of Y and let $u \in X \otimes Z$. We shall show that

$$\pi(u; X \otimes Y) \le \pi(u; X \otimes Z) \le \lambda\pi(u; X \otimes Y).$$

Let $\varepsilon > 0$. Then there exists a finite dimensional subspace M of X such that $\pi(u; M \otimes Y) \le (1+\varepsilon)\pi(u; X \otimes Y)$. Since this inequality is unchanged if M is replaced by a larger subspace of X, we may assume that there exists an isomorphism $T\colon M \to \ell_1^k$ such that $\|T\|\|T^{-1}\| \le \lambda$. Let J be the embedding of M into X and let $S = JT^{-1}$. Now $u = ST \otimes I(u)$ where I denotes the identity on Y, and so

$$\begin{aligned}
&\pi(u; X \otimes Z) = \pi(ST \otimes I(u); X \otimes Z) \le \|S\|\pi(T \otimes I(u); \ell_1^k \otimes Z) \\
&= \|S\|\pi(T \otimes I(u); \ell_1^k \otimes Y) \le \|S\|\|T\|\pi(u; M \otimes Y) < \lambda(1+\varepsilon)\pi(u; X \otimes Y).
\end{aligned}$$

Since this holds for every $\varepsilon > 0$, we obtain $\pi(u; X \otimes Z) \le \lambda\pi(u; X \otimes Y)$. □

We leave it as an exercise for the reader to show that X is a $\mathcal{L}_{1,1+\varepsilon}$-space for every $\varepsilon > 0$ if and only if the following is true: for every $\varepsilon > 0$, and every finite dimensional subspace M of X, there is a finite dimensional subspace, N, containing M, and an isomorphism $T\colon N \to \ell_1^k$ for some k, that satisfies $|\|Tx\| - \|x\|| \le \varepsilon\|x\|$ for every $x \in N$.

We show next that the space $L_1(\mu)$ is a $\mathcal{L}_{1,1+\varepsilon}$-space for every $\varepsilon > 0$. We shall make use of some elementary facts about bases in finite dimensional spaces. Let X be a finite dimensional normed space. Then X has an *Auerbach basis*, namely a basis $\{e_1, \ldots, e_n\}$ of unit vectors such that the coordinate funcionals $\{e_1^*, \ldots, e_n^*\}$ also have unit norm. Thus, if $x = \sum_{i=1}^n x_i e_i$, then $|x_i| \le \|x\|$ for every i. It follows that, if X is an n-dimensional subspace of any Banach space Z, then there is a projection of Z onto X of norm at most n. To define such a projection, let $\varphi_1, \ldots, \varphi_n$ be Hahn–Banach extensions of the coordinate functionals $e_1^*, \ldots, e_n^*$ to Z, and take $Tz = \sum_{i=1}^n \varphi_i(z)e_i$ for $z \in Z$.

Proposition 2.21. *$L_1(\mu)$ is a $\mathcal{L}_{1,1+\varepsilon}$-space for every $\varepsilon > 0$.*

Proof. Let M be a finite dimensional subspace of $L_1(\mu)$ and let $\varepsilon > 0$. Choose an Auerbach basis, $\{f_1, \dots, f_n\}$ for M. For each i, let g_i be an integrable simple function such that $\|f_i - g_i\|_1 < \varepsilon/n^2$. Then it is easy to see that the functions $g_1, \dots, g_n$ are linearly independent. Let M' be the subspace of $L_1(\mu)$ spanned by $g_1, \dots, g_n$. Let $S: M \to M'$ be the isomorphism given by $Sf_i = g_i$ for each i. If $f = \sum_{i=1}^n c_i f_i \in M$, then

$$\|f - Sf\|_1 \le \sum_{i=1}^n |c_i| \|f_i - g_i\|_1 \le \frac{\varepsilon}{n^2} \sum_{i=1}^n |c_i| \le \frac{\varepsilon}{n} \|f\|_1 . \qquad (*)$$

Now since g_i are measurable simple functions, there exist mutually disjoint measurable sets $E_1, \dots, E_k$, each having finite, positive measure, such that each g_i lies in the subspace N' of $L_1(\mu)$ spanned by the characteristic functions $\chi_{E_1}, \dots, \chi_{E_k}$. It is easy to see that N' is isometrically isomorphic to ℓ_1^k. By the remarks preceding the proposition, there is a subspace G of N' such that $N' = M' \oplus G$, and the projection of N' onto M' with kernel G has norm at most n. Let $N = M \oplus G$ and define an isomorphism $T: N \to N' = \ell_1^k$ as follows. If $f \in N$, then $f = h + k$, where $h \in M$ and $k \in G$. We define $Tf = Sh + k$. It is clear that T is an algebraic isomorphism. Furthermore, using $(*)$, we have

$$\big| \|f\|_1 - \|Tf\|_1 \big| \le \|f - Tf\|_1 = \|h - Sh\|_1 \le \frac{\varepsilon}{n} \|h\|_1 \le \varepsilon \|f\|_1 ,$$

since $\|h\|_1 \le n \|f\|_1$. □

The dual spaces $C(K)^*$ are also $\mathcal{L}_{1,\lambda}$-spaces for every $\lambda > 1$. We leave the proof of this fact as an exercise for the reader. It can be shown that a Banach space is an $\mathcal{L}_{1,\lambda}$-space for every $\lambda > 1$ if and only if it is an $L_1(\mu)$ for some μ. However, we shall not need to make use of this fact.

2.5 Rademacher Techniques

We have encountered two ways to compute the projective norm of a tensor u in $X \otimes Y$. On the one hand, we have the original definition:

$$\pi(u) = \inf \left\{ \sum_{i=1}^n \|x_i\| \|y_i\| : u = \sum_{i=1}^n x_i \otimes y_i \right\} .$$

This is complemented by the duality formula:

$$\pi(u) = \sup\{ |\langle u, B \rangle| : B \in \mathcal{B}(X \times Y), \|B\| \le 1 \} .$$

In order to use the first formula, we need to find the most efficient representation of u. In this section, we show how this can be done in certain special cases, using an averaging technique.

We illustrate this technique with a simple example. Consider the tensor $u = e_1 \otimes e_1 + e_2 \otimes e_2$ in $c_0 \otimes_\pi c_0$. Using this representation of u, we see that $\pi(u) \leq 2$. However, we can also express u as

$$u = e_1 \otimes e_1 + e_2 \otimes e_2 = \frac{1}{4}\Big[(e_1 + e_2) \otimes (e_1 + e_2) + (e_1 - e_2) \otimes (e_1 - e_2) + (-e_1 + e_2) \otimes (-e_1 + e_2) + (-e_1 - e_2) \otimes (-e_1 - e_2)\Big],$$

from which we obtain the estimate $\pi(u) \leq 1$. In fact, we have $\pi(u) = 1$, as we can see by applying the unit norm bilinear form $B(x, y) = x_1 y_1$ and using the duality formula.

This idea can be extended to a sum of n terms. It is not difficult to verify the identity

$$x_1 \otimes y_1 + \cdots + x_n \otimes y_n = \frac{1}{2^n} \sum_{\substack{\varepsilon_i = \pm 1 \\ 1 \leq i \leq n}} (\varepsilon_1 x_1 + \cdots + \varepsilon_n x_n) \otimes (\varepsilon_1 y_1 + \cdots + \varepsilon_n y_n). \quad (2.5)$$

Although the number of terms in the representation has increased from n to 2^n, this is compensated for by the factor 2^{-n}. This simple averaging technique has many applications. It can be formulated most elegantly in terms of the Rademacher functions, which we now describe.

The Rademacher functions are a sequence of functions on the interval $[0, 1]$, with values ± 1. The first function in the sequence is the constant function $r_1(t) = 1$. The second is obtained by bisecting the interval $[0, 1]$ and defining $r_2(t)$ to be 1 on $[0, 1/2)$ and -1 on $[1/2, 1]$. Once again these intervals are bisected and $r_3(t)$ takes the constant values 1 and -1 alternately on the intervals $[0, 1/4)$, $[1/4, 1/2)$, $[1/2, 3/4)$ and $[3/4, 1]$. In general, the function $r_{n=1}(t)$ takes the values 1 and -1 alternately on the 2^n subintervals $[0, 1/2^n), \ldots, [(2^n - 1)/2^n, 1]$. These functions are mutually orthogonal on $[0, 1]$:

$$\int_0^1 r_i(t) r_j(t)\, dt = \begin{cases} 1 & \text{if } i = j, \\ 0 & \text{if } i \neq j. \end{cases}$$

To see this, suppose that $i < j$. The function $r_i(t)$ is constant on each of 2^i subintervals of $[0, 1]$ and $r_j(t)$ takes the values 1 and -1 equally often on each of these subintervals. Therefore the integral is zero. On the other hand, if $i = j$, then $r_i(t) r_j(t) = r_i(t)^2 = 1$ for every t and so the integral is 1. Using the same argument, we can establish a more general orthogonality statement. If $k_1, \ldots, k_n$ are natural numbers, then

$$\int_0^1 r_1(t)^{k_1} \dots r_n(t)^{k_n}\, dt = \begin{cases} 1 & \text{if } k_1, \dots, k_n \text{ are all even,} \\ 0 & \text{otherwise.} \end{cases} \tag{2.6}$$

We now present the "averaging identity" in terms of the Rademacher functions. We point out that integrals such as those in (2.6) above are actually finite sums, since the interval $[0,1]$ can be partitioned into a finite number of subintervals in each of which the functions $r_j(t)$ that appear in the integral are constant.

Lemma 2.22 (Rademacher averaging). *Let E and F be vector spaces and let $x_1, \dots, x_n \in E$ and $y_1, \dots, y_n \in F$. Then*

$$\sum_{i=1}^n x_i \otimes y_i = \int_0^1 \left(\sum_{i=1}^n r_i(t) x_i \right) \otimes \left(\sum_{i=1}^n r_i(t) y_i \right) dt\,.$$

Proof. Expand the integrand and use the orthogonality property of the functions $r_i(t)$. □

The following example shows how Rademacher averaging can help us compute projective norms.

Example 2.23. *The tensor diagonal in $\ell_p \hat{\otimes}_\pi \ell_p$, $1 \le p \le \infty$.*

By the tensor diagonal we mean the closed subspace spanned by the tensors $e_n \otimes e_n$, where e_n are the standard unit basis vectors. We shall denote this subspace by D_p.

(a) The case $p = \infty$. Consider the tensor $u = \sum_{i=1}^n a_i e_i \otimes e_i = \sum_{i=1}^n (a_i e_i) \otimes e_i$. Using Rademacher averaging, we have

$$u = \int_0^1 \left(\sum_{i=1}^n a_i r_i(t) e_i \right) \otimes \left(\sum_{i=1}^n r_i(t) e_i \right) dt\,.$$

We may interpret this integral as a Bochner integral and hence

$$\pi(u) \le \sup_{0 \le t \le 1} \left\| \sum_{i=1}^n a_i r_i(t) e_i \right\|_\infty \left\| \sum_{i=1}^n r_i(t) e_i \right\|_\infty = \|(a_i)\|_\infty\,,$$

since $|r_i(t)| = 1$ for all values of i and t. Now choose an index j between 1 and n such that $\|(a_i)\|_\infty = |a_j|$ and consider the bilinear form on $\ell_\infty \times \ell_\infty$ given by $B(x,y) = \operatorname{sgn}(a_j) x_j y_j$. We have $\|B\| = 1$ and $\langle u, B \rangle = |a_j| = \|(a_i)\|_\infty$, which, by the duality formula, implies that $\pi(u) \ge \|(a_i)\|_\infty$. Therefore, $\pi(u) = \|(a_i)\|_\infty$. It follows that the mapping $\sum_{i=1}^n a_i e_i \in c_0 \mapsto \sum_{i=1}^n a_i e_i \otimes e_i \in D_\infty$ extends to an isometric isomorphism between these spaces. Therefore D_∞ may be identified with c_0.

(b) The case $2 < p < \infty$. Once again, we employ Rademacher averaging to produce a more efficient representation of the tensor $u = \sum_{i=1}^n a_i e_i \otimes e_i$, but now we distribute the coefficients in a different way:

$$u = \int_0^1 \Big(\sum_{i=1}^n \operatorname{sgn}(a_i)|a_i|^{1/2} r_i(t) e_i\Big) \otimes \Big(\sum_{i=1}^n |a_i|^{1/2} r_i(t) e_i\Big)\, dt\,.$$

This yields

$$\begin{aligned}\pi(u) &\le \sup_{0\le t\le 1} \Big\|\sum_{i=1}^n \operatorname{sgn}(a_i)|a_i|^{1/2} r_i(t) e_i\Big\|_p \Big\|\sum_{i=1}^n |a_i|^{1/2} r_i(t) e_i\Big\|_p \\ &= \Big(\sum_{i=1}^n |a_i|^{p/2}\Big)^{2/p} = \|(a_i)\|_{p/2}\,.\end{aligned}$$

To prove the reverse inequality, consider the bilinear form on $\ell_p \times \ell_p$ defined by $B(x,y) = \sum_{i=1}^n b_i x_i y_i$, where $b_i = \operatorname{sgn}(a_i)|a_i|^{p/2-1}$. We have $\langle u, B\rangle = \sum_{i=1}^n |a_i|^{p/2}$. In order to estimate the norm of B, we use Hölder's inequality. Thus

$$|\sum_{i=1}^n b_i x_i y_i| \le \|(b_i)\|_\alpha \|x\|_\beta \|y\|_\gamma\,,$$

where $0 \le \alpha, \gamma, \beta \le 1$ and $1/\alpha + 1/\beta + 1/\gamma = 1$. We take $\alpha = p/(p-2)$, and $\beta = \gamma = p$. This gives $\|B\| \le (\sum_{i=1}^n |a_i|^{p/2})^{(p-2)/p}$. Therefore

$$\sum_{i=1}^n |a_i|^{p/2} = |\langle u, B\rangle| \le \pi(u)\Big(\sum_{i=1}^n |a_i|^{p/2}\Big)^{(p-2)/p}\,,$$

from which we obtain $\|(a_i)\|_{p/2} \le \pi(u)$. Therefore $\pi(u) = \|(a_i)\|_{p/2}$. It follows that D_p is isometrically isomorphic to $\ell_{p/2}$.

(c) The case $1 \le p \le 2$. This case is argued exactly as in Example 2.10 and we find that D_p is isometrically isomorphic to ℓ_1 for these values of p.

To summarize, we have shown that the tensor diagonal D_p in $\ell_p \hat{\otimes}_\pi \ell_p$ satisfies

$$D_p = \begin{cases} \ell_1 & \text{if } 1 \le p \le 2\,, \\ \ell_{p/2} & \text{if } 2 < p < \infty\,, \\ c_0 & \text{if } p = \infty\,. \end{cases}$$

The Rademacher functions belong to the space $L_p[0,1]$ for every p. Our next result will enable us to identify the closed subspace of $L_p[0,1]$ generated by them. To do this, we must compute L_p-norms of the form

$$\Big\|\sum_{j=1}^n c_j r_j\Big\|_p\,,$$

where c_j are scalars. The case $p = 2$ is easy, since the Rademacher functions form an orthonormal sequence in $L_2[0,1]$. Therefore

$$\Big\|\sum_{j=1}^n c_j r_j\Big\|_2 = \|c\|_2\,, \tag{2.7}$$

where $\|c\|_2$ is the ℓ_2-norm of the vector $c = (c_1, \dots, c_n, 0, 0, \dots)$. It is a surprising fact that the L_p-norm of $\sum_{j=1}^n c_j r_j$ is equivalent to the ℓ_2-norm of the vector c for every value of p.

Theorem 2.24 (Khinchine's Inequality). *Let r_j be the Rademacher functions. For each $p \in [1, \infty)$ there are positive constants A_p and B_p such that*

$$A_p\|c\|_2 \leq \left\| \sum_{j=1}^n c_j r_j \right\|_p \leq B_p\|c\|_2$$

for every finite sequence $c = (c_1, \dots, c_n)$ of scalars.

Proof. We have already seen that $\|\sum_{j=1}^n c_j r_j\|_2 = \|c\|_2$ and so we may take $A_2 = B_2 = 1$. Now the norm of $L_p[0,1]$ is an increasing function of p and so it suffices to prove the B_p inequality for $p > 2$ and the A_p inequality for $p < 2$.

Step 1: The B_p inequality for $p > 2$ and $\mathbb{K} = \mathbb{R}$.
By the monotonicity of the L_p-norms, it is enough to prove the inequality when p is an even positive integer. Accordingly, let $p = 2m$. Using the Multinomial Theorem, we have

$$\begin{aligned}\left\| \sum_{j=1}^n c_j r_j \right\|_p^p &= \int_0^1 \left(\sum_{j=1}^n c_j r_j(t) \right)^{2m} dt \\ &= \sum_{k_1+\dots+k_n=2m} \frac{(2m)!}{k_1! \dots k_n!} c_1^{k_1} \dots c_n^{k_n} \int_0^1 r_1^{k_1}(t) \dots r_n^{k_n}(t)\, dt\,.\end{aligned}$$

The integral is zero unless all the k_j are even, say $k_j = 2m_j$. Now $k_1 + \dots + k_n = 2m$ if and only if $m_1 + \dots + m_n = m$ and so

$$\begin{aligned}\left\| \sum_{j=1}^n c_j r_j \right\|_p^p &= \sum_{m_1+\dots+m_n=m} \frac{(2m)!}{(2m_1)! \dots (2m_n)!} c_1^{2m_1} \dots c_n^{2m_n} \\ &\leq (2m)! \sum_{m_1+\dots+m_n=m} (c_1^2)_1^m \dots (c_n^2)_n^m \leq (2m)! \left(\sum_{j=1}^n c_j^2 \right)^m .\end{aligned}$$

Therefore

$$\left\| \sum_{j=1}^n c_j r_j \right\|_p \leq (p!)^{1/p}\|c\|_2\,,$$

when p is an even integer.

Step 2: The B_p inequality for $p > 2$ and $\mathbb{K} = \mathbb{C}$.
If $c_j = a_j + ib_j$ are complex numbers, then

$$\begin{aligned}\left\| \sum_{j=1}^n c_j r_j \right\|_p &\leq \left\| \sum_{j=1}^n a_j r_j \right\|_p + \left\| \sum_{j=1}^n b_j r_j \right\|_p \leq B_p(\|a\|_2 + \|b\|_2) \\ &\leq \sqrt{2} B_p(\|a\|_2^2 + \|b\|_2^2)^{1/2} = \sqrt{2} B_p\|c\|_2\,.\end{aligned}$$

Step 3: The A_p inequality for $1 \le p < 2$.
By the monotonicity of the L_p-norms, it suffices to prove the inequality in the case $p = 1$. Using Hölder's inequality with the conjugate exponents $3/2$ and 3 and the B_p inequality for $p = 4$, we have

$$\begin{aligned} \|c\|_2^2 &= \int_0^1 \left|\sum_{j=1}^n c_j r_j\right|^2 dt = \int_0^1 \left|\sum_{j=1}^n c_j r_j\right|^{2/3} \left|\sum_{j=1}^n c_j r_j\right|^{4/3} dt \\ &\le \left(\int_0^1 \left|\sum_{j=1}^n c_j r_j\right| dt\right)^{2/3} \left(\int_0^1 \left|\sum_{j=1}^n c_j r_j\right|^4 dt\right)^{1/3} \\ &\le \left\|\sum_{j=1}^n c_j r_j\right\|_1^{2/3} B_4^{4/3} \|c\|_2^{4/3} . \end{aligned}$$

It follows that

$$\left\|\sum_{j=1}^n c_j r_j\right\|_1 \ge B_4^{-2} \|c\|_2$$

and so we have proved the A_1 inequality with $A_1 = B_4^{-2}$. □

We should point out that the proof we have given here does not by any means give the best values for the constants A_p and B_p.

Khinchine's inequality shows that the closed subspace of $L_p[0,1]$ generated by the Rademacher functions is isomorphic to ℓ_2. For values of p greater than 1 we can say a little more about this subspace:

Theorem 2.25. *Let $1 \le p < \infty$. The closed subspace R_p of $L_p[0,1]$ generated by the Rademacher functions is isomorphic to ℓ_2. If $p > 1$ then R_p is complemented in $L_p[0,1]$.*

Proof. It follows from Khinchine's inequality that the mapping $\sum_{j=1}^n c_j e_j \mapsto \sum_{j=1}^n c_j r_j$ extends to an isomorphism from ℓ_2 onto R_p.

Now suppose that $p \ge 2$. Then $L_p[0,1] \subset L_2[0,1]$ and so, for every f in $L_p[0,1]$, the "Rademacher coefficients",

$$\langle f, r_n \rangle = \int_0^1 f(t) r_n(t)\, dt ,$$

of f are square summable. We claim that the required projection of $L_p[0,1]$ onto R_p is given by the formula

$$P_p f = \sum_{n=1}^{\infty} \langle f, r_n \rangle r_n .$$

First, we use the Cauchy criterion to show that this series converges in $L_p[0,1]$. We have

$$\left\|\sum_{j=n}^{m}\langle f,r_j\rangle r_j\right\|_p \le B_p\left(\sum_{j=n}^{m}|\langle f,r_j\rangle|^2\right)^{1/2},$$

and, since $(\langle f,r_n\rangle) \in \ell_2$, it follows that $\sum_{n=1}^{\infty}\langle f,r_n\rangle r_n$ converges to an element of $L_p[0,1]$. Therefore, the series defining P_pf converges in the space $L_p[0,1]$. To see that P_p is bounded, we use the Khinchine inequality again, together with Bessel's inequality. For every n, we have

$$\left\|\sum_{j=1}^{n}\langle f,r_j\rangle r_j\right\|_p \le B_p\left(\sum_{j=1}^{n}|\langle f,r_j\rangle|^2\right)^{1/2} = B_p\left\|\sum_{j=1}^{n}\langle f,r_j\rangle r_j\right\|_2$$
$$\le B_p\|f\|_2 \le B_p\|f\|_p\,.$$

Letting $n \to \infty$ we get

$$\|P_pf\|_p \le B_p\|f\|_p\,.$$

Now it is easy to see that P_p is a projection and so it only remains to identify the range of P_p. The definition of P_p shows that $P_p(L_p[0,1]) \subset R_p$. On the other hand, $P_pr_n = r_n$ for every n implies that $R_p \subset P_p(L_p[0,1])$. Therefore $P_p(L_p[0,1]) = R_p$.

Now let $1 < p < 2$ and let q be the conjugate index of p. In this case, we define $P_p = P_q^*$, where P_q is the projection of $L_q[0,1]$ onto R_q constructed above. Now P_p, being the adjoint of a bounded projection, is itself a bounded projection. Thus we have only to show that the range of P_p is R_p. First, we show that $P_pr_n = r_n$ for every n, from which it follows that $R_p \subset P_p(L_p[0,1])$. Let $g \in L_q[0,1]$. Then

$$\langle g, P_pr_n\rangle = \langle P_qg, r_n\rangle = \left\langle \sum_{k=1}^{\infty}\langle g, r_k\rangle r_k, r_n\right\rangle = \langle g, r_n\rangle$$

and so $P_pr_n = r_n$. We claim that the series $\sum_{n=1}^{\infty}\langle f,r_n\rangle r_n$ converges weakly to P_pf for every $f \in L_p[0,1]$. Since the weak and norm closures of a subspace are the same, this implies that $P_p(L_p[0,1]) \subset R_p$ and the proof is complete. To establish our claim, fix $g \in L_q[0,1]$. Then

$$\left\langle g, \sum_{j=1}^{n}\langle f,r_j\rangle r_j\right\rangle = \sum_{j=1}^{n}\langle f,r_j\rangle\langle g,r_j\rangle$$
$$= \left\langle \sum_{j=1}^{n}\langle g,r_j\rangle r_j, f\right\rangle \to \langle P_qg, f\rangle = \langle g, P_pf\rangle$$

as $n \to \infty$. Therefore P_pf is the sum of the weakly convergent series $\sum_{n=1}^{\infty}\langle f,r_n\rangle r_n$. □

It can be shown that the subspace R_1 of $L[0,1]$ is not complemented (see Exercise 6.11).

We can deduce some interesting structural information about projective tensor products from this proposition. We know that the projective tensor product respects complemented subspaces isomorphically and we have seen in Example 2.10 that $\ell_2\hat{\otimes}_\pi\ell_2$ contains a complemented isometric copy of ℓ_1. Thus we have

Corollary 2.26. *Let $1 < p_1, p_2 < \infty$. Then $L_{p_1}[0,1]\hat{\otimes}_\pi L_{p_2}[0,1]$ contains a complemented subspace that is isomorphic to ℓ_1.*

Once again, we see that the projective tensor product has some "bad" properties – the tensor product of reflexive spaces fails dramatically to be reflexive.

2.6 Nuclear Bilinear Forms and Operators

We have seen in Chapter 1 that tensors can be viewed as bilinear forms by means of the canonical linear mapping that associates with a tensor $u = \sum_{i=1}^n \varphi_i \otimes \psi_i \in X^\sharp \otimes Y^\sharp$ the bilinear form $B_u \in B(X \times Y)$ given by the formula $B_u(x,y) = \sum_{i=1}^n \varphi_i(x)\psi_i(y)$. Now if X and Y are Banach spaces, the elements of the tensor product $X^* \otimes Y^*$ define bounded bilinear forms on $X \times Y$ and we obtain an injective operator of unit norm from $X^* \otimes_\pi Y^*$ into $\mathcal{B}(X \times Y)$. Taking the extension of this operator to the completion gives an operator

$$J\colon X^*\hat{\otimes}_\pi Y^* \to \mathcal{B}(X \times Y)$$

of unit norm. A bilinear form on $X \times Y$ is said to be *nuclear* if it lies in the range of this operator. Bearing in mind the explicit description we have found in Section 1 for the elements of a completed tensor product, we see that a bilinear form B is nuclear if and only if there exist bounded sequences (φ_n) and (ψ_n) in X^* and Y^* respectively, satisfying $\sum_{n=1}^\infty \|\varphi_n\|\|\psi_n\| < \infty$, such that

$$B(x,y) = \sum_{n=1}^\infty \varphi_n(x)\psi_n(y)$$

for every $x \in X, y \in Y$. We shall refer to an expression of the form $\sum_{n=1}^\infty \varphi_n \otimes \psi_n$ as a *nuclear representation* of B. The *nuclear norm* of B is defined to be

$$\|B\|_N = \inf\left\{\sum_{n=1}^\infty \|\varphi_n\|\|\psi_n\| : B(x,y) = \sum_{n=1}^\infty \varphi_n(x)\psi_n(y)\right\},$$

the infimum being taken over all the nuclear representations of B. It is easy to see that

$$\|B\| \le \|B\|_N$$

and it follows that the nuclear norm is indeed a norm. We shall denote by $\mathcal{B}_N(X \times Y)$ the space of nuclear bilinear forms with the nuclear norm. We leave it as an exercise to the reader to verify that $\mathcal{B}_N(X \times Y)$ is a Banach space.

It seems at first glance that $\mathcal{B}_N(X \times Y)$ is just $X^*\hat{\otimes}_\pi Y^*$ in disguise. A closer examination reveals that this is not necessarily the case. The difficulty is that the canonical mapping

$$J\colon X^*\hat{\otimes}_\pi Y^* \to \mathcal{B}_N(X \times Y)$$

might fail to be injective. Let $u \in X^*\hat{\otimes}_\pi Y^*$ and let $\sum_{n=1}^{\infty} \varphi_n \otimes \psi_n$ be a representation of u. Now, if the corresponding bilinear form B is zero, then $B(x,y) = \sum_{n=1}^{\infty} \varphi_n(x)\psi_n(y) = 0$ for every $x \in X$, $y \in Y$. But in order to deduce that $u = 0$, we must have $\langle u, A\rangle = \sum_{n=1}^{\infty} A(\varphi_n, \psi_n) = 0$ for every $A \in \mathcal{B}(X^* \times Y^*) = (X^*\hat{\otimes}_\pi Y^*)^*$. In general, these conditions are not equivalent. The "obstruction" to the identity $\mathcal{B}_N(X \times Y) = X^*\hat{\otimes}_\pi Y^*$ is the kernel of J:

$$\ker J = \{u \in X^*\hat{\otimes}_\pi Y^* : B_u(x,y) = 0 \quad \forall x \in X, y \in Y\}.$$

Looking again at the definition of the nuclear norm, we see that it is the quotient norm of $X^*\hat{\otimes}_\pi Y^* / \ker J$ and so the best we can say in general is that

$$\mathcal{B}_N(X \times Y) = X^*\hat{\otimes}_\pi Y^* / \ker J\,.$$

We shall investigate this problem in greater detail in Chapter 4.

For now, we shall be content to note one example in which the space of nuclear bilinear forms coincides with the projective tensor product.

Example 2.27. *Nuclear bilinear forms on* c_0.

We have $c_0^*\hat{\otimes}_\pi c_0^* = \ell_1\hat{\otimes}_\pi\ell_1 = \ell_1(\ell_1)$. Now the space $\ell_1(\ell_1)$ is isometrically isomorphic to the space $\ell_1(\mathbb{N}\times\mathbb{N})$ of summable families of scalars indexed by $\mathbb{N} \times \mathbb{N}$. Let $\{e_{nm}\}$ be the standard unit vector basis for this space, so that a typical element of $\ell_1(\mathbb{N} \times \mathbb{N})$ has the form $a = (a_{nm}) = \sum_{n,m} a_{nm} e_{nm}$, with $\|a\|_1 = \sum_{n,m} |a_{nm}|$. We now have an identification of $c_0^*\hat{\otimes}_\pi c_0^*$ with $\ell_1(\mathbb{N}\times\mathbb{N})$, under which the elementary tensor $e_n^* \otimes e_m^*$ corresponds to the basis vector e_{nm}. It follows that every $u \in c_0^*\hat{\otimes}_\pi c_0^*$ has the form $u = \sum_{n,m} a_{nm} e_n^* \otimes e_m^*$, with $\pi(u) = \sum_{n,m} |a_{nm}|$. Therefore the corresponding nuclear bilinear form is given by

$$B_u(x,y) = \sum_{n,m} a_{nm} x_n y_m\,.$$

Now, if $B_u = 0$, then by taking $x = e_n$ and $y = e_m$ we get $a_{nm} = 0$ for every n,m and hence $u = 0$. So, in this case, the kernel of the canonical mapping is trivial and we have

$$\mathcal{B}_N(c_0 \times c_0) = \ell_1\hat{\otimes}_\pi\ell_1 = \ell_1(\mathbb{N} \times \mathbb{N})\,.$$

We can also look at nuclearity from the point of view of operators. There is a canonical operator $J\colon X^*\hat{\otimes}_\pi Y \to \mathcal{L}(X,Y)$ of unit norm that associates with the tensor $u = \sum_{n=1}^\infty \varphi_n \otimes y_n$ the operator $L_u\colon X \to Y$, defined by $L_u(x) = \sum_{n=1}^\infty \varphi_n(x)y_n$. The operators that arise in this way are called *nuclear.* The nuclear operators form a Banach space, which we shall denote by $\mathcal{N}(X,Y)$, or simply $\mathcal{N}(X)$ when $Y = X$, with the *nuclear norm*:

$$\|T\|_N = \inf\left\{\sum_{n=1}^\infty \|\varphi_n\|\|y_n\| : T(x) = \sum_{n=1}^\infty \varphi_n(x)y_n\right\},$$

the infimum being taken over all the representations of T of the form $T(x) = \sum_{n=1}^\infty \varphi_n(x)y_n$, where (φ_n) and (y_n) are bounded sequences in X^* and Y respectively such that $\sum_{n=1}^\infty \|\varphi_n\|\|y_n\| < \infty$. As for nuclear bilinear forms, we can identify the space $\mathcal{N}(X,Y)$ with a quotient space of the projective tensor product:

$$\mathcal{N}(X,Y) = X^*\hat{\otimes}_\pi Y/\ker J\,.$$

We leave it as an exercise to the reader to verify that, if an operator T is nuclear, then so is its adjoint. Furthermore, we have $\|T^*\|_N \le \|T\|_N$. The converse of this statement, and the question of whether the nuclear norms of T and T^* coincide, is more difficult and will be dealt with in Chapter 4.

The proof of the following proposition is routine and will be left to the reader.

Proposition 2.28. *Let $S\colon X \to Y$ be a nuclear operator and let $T \in \mathcal{L}(W,X)$ and $R \in \mathcal{L}(Y,Z)$. Then the operator RST is nuclear and*

$$\|RST\|_N \le \|R\|\|S\|_N\|T\|\,.$$

As is the case with bilinear forms, we have

$$\|T\| \le \|T\|_N$$

for every nuclear operator. The next result shows that the nuclear norm can be much bigger than the operator norm.

Proposition 2.29. *Let X have dimension n. Then the identity operator on X has nuclear norm n.*

Proof. We note first that the tensor product $X^* \otimes_\pi X$ is complete and can be identified with $\mathcal{N}(X)$. Choose an Auerbach basis $\{e_1,\dots,e_n\}$ for X with associated coordinate functionals $e_1^*,\dots,e_n^*$. Thus we have $\|e_i\| = \|e_i^*\| = 1$ for each i. Then $x = \sum_{i=1}^n e_i^*(x)e_i$ for every $x \in X$ and so the identity operator, I, is given by the tensor $\sum_{i=1}^n e_i^* \otimes e_i$. Therefore

$$\|I\|_N = \pi\left(\sum_{i=1}^n e_i^* \otimes e_i\right) \le n$$

by the definition of the projective norm. Now consider the trace functional, tr_X on $X^* \otimes_\pi X$. We have $\langle \varphi \otimes x, \mathrm{tr}_X \rangle = \varphi(x)$, and tr_X has norm one. Hence

$$n = \left\langle \sum_{i=1}^{n} e_i^* \otimes e_i, \mathrm{tr}_X \right\rangle \leq \pi\left(\sum_{i=1}^{n} e_i^* \otimes e_i \right) = \|I\|_N .$$

Therefore $\|I\|_N = n$. □

In infinite dimensions, the class of nuclear operators is quite small. For example, we can see from the definition that every nuclear operator is the limit in operator norm of a sequence of finite rank operators. Thus, in particular, every nuclear operator is compact.

We have seen that the nuclear bilinear forms on $c_0 \times c_0$ have a simple description. We now look at nuclear operators on c_0.

Example 2.30. *Nuclear operators on* c_0.

We recall that $c_0^* \hat{\otimes}_\pi X = \ell_1 \hat{\otimes}_\pi X$ can be identified with the space $\ell_1(X)$. Under this identification, the canonical mapping $J\colon c_0^* \hat{\otimes}_\pi X = \ell_1(X) \to \mathcal{N}(c_0, X)$ associates with the element $x = (x_n)$ of $\ell_1(X)$ the nuclear operator given by $T(a) = \sum_{n=1}^{\infty} a_n x_n$. It follows easily that J is injective. Hence we have

$$\mathcal{N}(c_0, X) = \ell_1 \hat{\otimes}_\pi X = \ell_1(X) ,$$

and, for every nuclear operator T,

$$\|T\|_N = \sum_{n=1}^{\infty} \|Te_n\| .$$

We conclude with some remarks about finite dimensional spaces. If X and Y are finite dimensional, then every operator from X into Y is nuclear, and, of course, the tensor product $X^* \otimes_\pi Y$ is complete. Thus, $\mathcal{N}(X,Y) = X^* \otimes_\pi Y$. In the language of trace duality, we see that the nuclear norm and the operator norm are in duality; the spaces $\mathcal{N}(X,Y)$ and $\mathcal{L}(Y,X)$ are dual, and for every $T \in \mathcal{L}(X,Y)$,

$$\|T\|_N = \sup\{ |\mathrm{tr}_X\, ST| : S \in \mathcal{L}(Y,X),\ \|S\| \leq 1 \} .$$

2.7 Exercises

Exercise 2.1. Show that $X \hat{\otimes}_\pi Y$ contains norm one complemented isometric copies of X and Y.

Exercise 2.2. Show that $X \hat{\otimes}_\pi Y$ is isometrically isomorphic to $Y \hat{\otimes}_\pi X$.

Exercise 2.3. Show that the Banach spaces $(X\hat{\otimes}_\pi Y)\hat{\otimes}_\pi Z$ and $X\hat{\otimes}_\pi(Y\hat{\otimes}_\pi Z)$ are isometrically isomorphic.

Exercise 2.4. Show that $X \otimes_\pi Y$ is complete if either X or Y is finite dimensional.

Exercise 2.5. Show that $X \otimes_\pi Y$ is never complete if X and Y are infinite dimensional.

Exercise 2.6. Show that $\ell_1\hat{\otimes}_\pi\ell_1$ is isometrically isomorphic to ℓ_1.

Exercise 2.7. Let X be a closed subspace of $C[0,1]$ that is isometrically isomorphic to ℓ_1. Show that $X \otimes_\pi c_0$ is *not* a subspace of $C[0,1] \otimes_\pi c_0$. (Hint: every operator from a $C(K)$ space into an $L_1(\mu)$ space is weakly compact.)

Exercise 2.8. Show that $L_1(\mu)\hat{\otimes}_\pi L_1(\nu)$ is isometrically isomorphic to the space $L_1(\mu \times \nu)$, where $\mu \times \nu$ is the product measure.

Exercise 2.9. In each of the following cases, determine whether the function $f\colon [0,1] \to X$ is Bochner integrable with respect to Lebesgue measure, m. If f is Bochner integrable, calculate the Bochner integral $\int_0^1 f\,dm$.
(a) $X = \ell_\infty$ and $f(t) = (r_n(t))$, where $r_n(t)$ are the Rademacher functions.
(b) $X = \ell_\infty$ and $f(t) = (n^{-1}r_n(t))$.
(c) $X = L_1[0,1]$ and $f(t) = \chi_{[0,t]}$.
(d) $X = L_\infty[0,1]$ and $f(t) = \chi_{[0,1]}$.

Exercise 2.10. Find a Banach space X and a function $f\colon [0,1] \to X$ such that $\|f\|$ is integrable with respect to Lebesgue measure, but f is not Bochner integrable.

Exercise 2.11. Show that $C(K)^*$ is a $\mathcal{L}_{1,\lambda}$-space for every $\lambda > 1$.

Exercise 2.12. Show that the sequence of Rademacher functions is not an orthonormal basis for $L_2[0,1]$.

Exercise 2.13. Show that the normed spaces $\mathcal{B}_N(X \times Y)$ and $\mathcal{N}(X,Y)$ are complete.

Exercise 2.14. Find an example of a compact operator that is not nuclear.

Exercise 2.15. (a) Let $T\colon X \to Y$ be a nuclear operator. Show that the adjoint operator $T^*\colon Y^* \to X^*$ nuclear and $\|T^*\|_N \leq \|T\|_N$.

(b) Let T be an operator whose adjoint T^* is nuclear. Why is it not possible to deduce that T is nuclear?

Exercise 2.16. An operator $T\colon H \to H$ on a Hilbert space is said to belong to the trace class if $\sum_{n=1}^\infty |\langle Te_n, e_n\rangle| < \infty$ for every orthonormal sequence $\{e_n\}$ in H.

(a) Show that every nuclear operator on H belongs to the trace class.

(b) Let T be defined on the *real* space ℓ_2 by $Tx = (-x_2, x_1, -x_4, x_3, \ldots)$. Show that T belongs to the trace class but is not nuclear.

3. The Injective Tensor Product

In this chapter we study the injective norm for tensor products. The injective tensor product gives a representation of Banach spaces of continuous vector valued functions and injective tensor products with $L_1(\mu)$ spaces provide an introduction to the Pettis integral. The duality theory of injective tensor products leads to the introduction of the important classes of integral bilinear forms and operators.

3.1 The Injective Norm

We shall now take a different approach to norming the tensor product, $X \otimes Y$, of two Banach spaces. We recall from Chapter 1 that the elements of the tensor product can be viewed as bilinear forms on the product $X^\sharp \times Y^\sharp$ of the algebraic duals. If $\sum_{i=1}^n x_i \otimes y_i$ is any representation of the tensor u, the associated bilinear form is given by

$$B_u(\varphi, \psi) = \sum_{i=1}^{n} \varphi(x_i)\psi(y_i)\,.$$

Now the restriction of B_u to the product $X^* \times Y^*$ of the dual spaces is bounded and so we have a canonical algebraic embedding of $X \otimes Y$ into $\mathcal{B}(X^* \times Y^*)$. The *injective norm* on $X \otimes Y$ is the norm induced by this embedding. We shall denote the injective norm of $u \in X \otimes Y$ by $\varepsilon(u)$, or by $\varepsilon_{X,Y}(u)$, or $\varepsilon(u; X \otimes Y)$, if it is necessary to specify the component spaces in the tensor product. Thus we have

$$\varepsilon(u) = \sup\left\{\left|\sum_{i=1}^{n} \varphi(x_i)\psi(y_i)\right| : \varphi \in B_{X^*}, \psi \in B_{Y^*}\right\}, \tag{3.1}$$

where $\sum_{i=1}^n x_i \otimes y_i$ is any representation of u. We may also view the elements of $X \otimes Y$ as operators, either from X^* into Y, or from Y^* into X. Thus, the operators $L_u \colon X^* \to Y$ and $R_u \colon Y^* \to X$ are given by $L_u\varphi = \sum_{i=1}^n \varphi(x_i)y_i$ and $R_u\psi = \sum_{i=1}^n \psi(y_i)x_i$, and these operators have the same norms as the bilinear form B_u. This gives two more formulas for the injective norm:

$$\begin{aligned}\varepsilon(u) &= \sup\left\{\left\|\sum_{i=1}^{n}\varphi(x_i)y_i\right\| : \varphi \in B_{X^*}\right\} \\ &= \sup\left\{\left\|\sum_{i=1}^{n}\psi(y_i)x_i\right\| : \psi \in B_{Y^*}\right\}.\end{aligned} \tag{3.2}$$

Unless there is a danger of ambiguity, we shall from now on use the same symbol for the tensor u and the bilinear form B_u, or either of the operators associated with u as described above.

A simple but invaluable observation is that we may replace the balls B_{X^*} and B_{Y^*} in any of these formulas by norming sets: a subset A of B_{X^*} is said to be a *norming set* if we have $\|x\| = \sup\{|\varphi(x)| : \varphi \in A\}$ for every $x \in X$. For example, the closed unit ball of X is a norming set in the bidual X^{**}, and so, if $u = \sum_{i=1}^{n} \varphi_i \otimes \psi_i \in X^* \otimes Y^*$, then

$$\varepsilon(u) = \sup\left\{\left|\sum_{i=1}^{n}\varphi_i(x)\psi_i(y)\right| : x \in B_X, y \in B_Y\right\}, \tag{3.3}$$

with similar variations on (3.2).

We denote by $X \otimes_\varepsilon Y$ the tensor product $X \otimes Y$ with the injective norm. The completion, denoted by $X\hat{\otimes}_\varepsilon Y$, is called the *injective tensor product* of X and Y. Unlike the projective tensor product, there is no general representation of the elements of the completed tensor product $X\hat{\otimes}_\varepsilon Y$. However, since $X \otimes_\varepsilon Y$ is a subspace of $\mathcal{B}(X^* \times Y^*)$, the completion, $X\hat{\otimes}_\varepsilon Y$ is simply the closure in $\mathcal{B}(X^* \times Y^*)$, or equivalently, in $\mathcal{L}(X^*, Y)$, or in $\mathcal{L}(Y^*, X)$. Thus the injective tensor product $X\hat{\otimes}_\varepsilon Y$ may be considered as a subspace of the space $\mathcal{B}(X^* \times Y^*)$ of bounded bilinear forms, or of either of the spaces of operators $\mathcal{L}(X^*, Y)$ or $\mathcal{L}(Y^*, X)$. And, in the special case in which one of the component spaces is a dual, $X^*\hat{\otimes}_\varepsilon Y$ is a subspace of $\mathcal{L}(X, Y)$. The operators from X into Y that arise in this way, as limits in the operator norm of sequences of finite rank operators, are known as *approximable operators*.

We summarize some useful embeddings of the injective tensor product into spaces of bilinear forms or operators:

$$\begin{aligned}X\hat{\otimes}_\varepsilon Y &\subset \mathcal{B}(X^* \times Y^*), \quad \mathcal{L}(X^*, Y), \quad \text{or} \quad \mathcal{L}(Y^*, X) \\ X^*\hat{\otimes}_\varepsilon Y &\subset \mathcal{B}(X \times Y^*), \quad \mathcal{L}(Y^*, X^*), \quad \text{or} \quad \mathcal{L}(X, Y) \\ X^*\hat{\otimes}_\varepsilon Y^* &\subset \mathcal{B}(X \times Y), \quad \mathcal{L}(X, Y^*), \quad \text{or} \quad \mathcal{L}(Y, X^*).\end{aligned} \tag{3.4}$$

We now record some of the elementary properties of the injective norm. The proofs are easy consequences of the definition and are left to the reader.

Proposition 3.1. *Let X and Y be Banach spaces.*

(a) $\varepsilon(u) \le \pi(u)$ *for every* $u \in X \otimes Y$.
(b) $\varepsilon(x \otimes y) = \|x\|\|y\|$ *for every* $x \in X$, $y \in Y$.

(c) *If* $\varphi \in X^*$, $\psi \in Y^*$, *then* $\varphi \otimes \psi$ *is a bounded linear functional on* $X\hat{\otimes}_\varepsilon Y$ *and* $\|\varphi \otimes \psi\| = \|\varphi\|\|\psi\|$.

We now consider the tensor product of operators.

Proposition 3.2. *Let* $S\colon X \to W$ *and* $T\colon Y \to Z$ *be operators. Then there exists a unique operator* $S\otimes_\varepsilon T\colon X\hat{\otimes}_\varepsilon Y \to W\hat{\otimes}_\varepsilon Z$ *such that* $(S\otimes_\varepsilon T)(x\otimes y) = (Sx)\otimes(Ty)$ *for every* $x \in X$, $y \in Y$. *Furthermore,* $\|S\otimes_\varepsilon T\| = \|S\|\|T\|$.

Proof. Let $S\otimes T\colon X\otimes Y \to W\otimes Z$ be the tensor product operator. If $u = \sum_{i=1}^n x_i \otimes y_i \in X\otimes Y$, then

$$\begin{aligned}
\varepsilon_{W,Z}(S\otimes T(u)) &= \sup\left\{\left|\sum_{i=1}^n \varphi(Sx_i)\psi(Ty_i)\right| : \varphi \in B_{W^*}, \psi \in B_{Z^*}\right\} \\
&= \sup\left\{\left|\sum_{i=1}^n (S^*\varphi)x_i(T^*\psi)y_i\right| : \varphi \in B_{W^*}, \psi \in B_{Z^*}\right\} \\
&\le \|S^*\|\|T^*\|\varepsilon_{X,Y}(u) = \|S\|\|T\|\varepsilon_{X,Y}(u)\,.
\end{aligned}$$

Therefore $S\otimes T$ is bounded for the injective norms, and has norm at most $\|S\|\|T\|$. On the other hand, for any $\varepsilon > 0$, we may choose $x \in B_X$ and $y \in B_Y$ such that $\|Sx\| \ge (1-\varepsilon)\|S\|$ and $\|Ty\| \ge (1-\varepsilon)\|T\|$. Then $\varepsilon(x\otimes y) \le 1$ and $\varepsilon((S\otimes T))(x\otimes y) \ge (1-\varepsilon)^2\|S\|\|T\|$. It follows that $\|S\otimes T\| \ge \|S\|\|T\|$ and therefore $\|S\otimes T\| = \|S\|\|T\|$. Now the operator $S\otimes T$ has a unique extension, with the same norm, to the completed tensor product. We define $S\otimes_\varepsilon T$ to be this extension.

□

It follows immediately from the formulas (3.2) that the injective tensor product respects subspaces, so that, if X is a closed subspace of W and Y is a closed subspace of Z, then $X\hat{\otimes}_\varepsilon Y$ is a closed subspace of $W\hat{\otimes}_\varepsilon Z$. However, the injective tensor product does not, in general, respect quotients; if X is a quotient space of W, then $X\hat{\otimes}_\varepsilon Y$ need not be a quotient of $W\hat{\otimes}_\varepsilon Y$. We refer to the exercises for further details on this.

We now look at some examples.

Example 3.3. *The injective tensor product* $c_0\hat{\otimes}_\varepsilon X$.

We shall show that $c_0\hat{\otimes}_\varepsilon X$ can be identified with the Banach space $c_0(X)$ of sequences in X that converge to zero, where this space carries the norm $\|(x_n)\| = \sup_n \|x_n\|$. Consider the canonical mapping $J\colon c_0\otimes X \to c_0(X)$ that takes the tensor $u = \sum_{i=1}^n a_i \otimes x_i$ to the X-valued sequence $Ju = (\sum_{i=1}^n a_{ik}x_i)_k$. To show that Ju is, in fact, an element of $c_0(X)$, we observe that

$$\left\|\sum_{i=1}^n a_{ik}x_i\right\| \le \max_{1\le i\le n} |a_{ik}| \sum_{i=1}^n \|x_i\| \to 0$$

as $k \to \infty$. Now we compute the norm of Ju. Using the duality between c_0 and ℓ_1, we have

$$\begin{aligned}\|Ju\| &= \sup_k \left\|\sum_{i=1}^n a_{ik}x_i\right\| = \sup_k \sup_{\varphi\in B_{X^*}} \left|\sum_{i=1}^n a_{ik}\varphi(x_i)\right| \\ &= \sup_{\varphi\in B_{X^*}} \sup_k \left|\sum_{i=1}^n a_{ik}\varphi(x_i)\right| = \sup_{\varphi\in B_{X^*}} \sup_{b\in B_{\ell_1}} \left|\sum_{k=1}^\infty\sum_{i=1}^n b_k a_{ik}\varphi(x_i)\right| \\ &= \sup_{b\in B_{\ell_1}} \sup_{\varphi\in B_{X^*}} \left|\sum_{i=1}^n b(a_i)\varphi(x_i)\right| = \varepsilon(u)\,,\end{aligned}$$

by (3.1). It follows that J extends to an isometry from $c_0\hat{\otimes}_\varepsilon X$ into $c_0(X)$. But it is easy to see that $J(c_0 \otimes X)$ is dense in $c_0(X)$ and therefore J is an isometric isomorphism.

Example 3.4. *The injective tensor product* $\ell_1\hat{\otimes}_\varepsilon X$.

Let $\ell_1[X]$ denote the space of all unconditionally summable sequences in X, in other words, sequences (x_n) such that the series $\sum x_n$ converges in X, regardless of the order of the terms. We shall make use of the Bounded Multiplier Test, which states that a sequence (x_n) belongs to $\ell_1[X]$ if and only if the series $\sum b_n x_n$ converges in X for every bounded sequence of scalars (b_n). (We refer the reader who is not familiar with this material to Appendix B for further details.) Now, if (x_n) is unconditionally summable, we may define a linear mapping $T\colon X^* \to \ell_1$ by the formula $T\varphi = (\varphi x_n)$, and an application of the Closed Graph Theorem shows that T is bounded. Furthermore, we see from the Bounded Multiplier Test that the adjoint operator T^* takes its values in X; we have $T^*b = \sum b_n x_n$ for every $b = (b_n) \in \ell_\infty$. It follows that T is continuous for the weak* topology on X^* and the weak topology on ℓ_1. Therefore, by Alaoglu's Theorem, T maps the weak* compact closed unit ball of X^* into a weakly compact subset of ℓ_1. But, by the Schur property of ℓ_1 (see Appendix C), every weakly compact set is compact in the norm topology. We may conclude that the operator $T\colon X^* \to \ell_1$ is compact, as is the adjoint $T^*: \ell_\infty \to X$. Now this association of each element of $\ell_1[X]$ with an operator suggests a natural norm for this space. Bearing in mind the fact that an operator has the same norm as its adjoint, we have two formulas for the norm of an element (x_n) of $\ell_1[X]$:

$$\|(x_n)\| = \sup\left\{\sum_{n=1}^\infty |\varphi(x_n)| : \varphi \in B_{X^*}\right\} = \sup\left\{\left\|\sum_{n=1}^\infty b_n x_n\right\| : b \in B_{\ell_\infty}\right\}.$$

It is a straightforward exercise, which we leave to the reader, to use the second of these formulas to show that $\ell_1[X]$ is complete in this norm. Furthermore, if we fix an element (x_n) of $\ell_1[X]$, then the compactness of the operator T implies that $\sup\{\sum_{k>n} |\varphi(x_k)| : \varphi \in B_{X^*}\} \to 0$ as $n \to \infty$. In other words, the sequence whose nth term is $(x_1,\ldots,x_n,0,0,\ldots)$ converges to (x_k) in $\ell_1[X]$.

Now let us show that $\ell_1[X]$ is isometrically isomorphic to the injective tensor product $\ell_1\hat{\otimes}_\varepsilon X$. We begin with the canonical mapping $J\colon \ell_1\otimes X\to\ell_1[X]$ which maps the tensor $u=\sum_{i=1}^n a_i x_i$ to the sequence $Ju=(\sum_{i=1}^n a_{ik}x_i)_k$. We compute the norm of Ju:

$$\begin{aligned}\|Ju\| &= \sup_{\varphi\in B_{X^*}}\sum_{k=1}^{\infty}\left|\varphi\Big(\sum_{i=1}^n a_{ik}x_i\Big)\right| = \sup_{\varphi\in B_{X^*}}\sup_{b\in B_{\ell_\infty}}\left|\sum_{k=1}^{\infty} b_k\Big(\sum_{i=1}^n a_{ik}\varphi(x_i)\Big)\right| \\ &= \sup_{\varphi\in B_{X^*}}\sup_{b\in B_{\ell_\infty}}\left|\sum_{i=1}^n b(a_i)\varphi(x_i)\right| = \varepsilon(u).\end{aligned}$$

Therefore J extends to an isometry of $\ell_1\hat{\otimes}_\varepsilon X$ into $\ell_1[X]$. But we have seen that the finite sequences are dense in $\ell_1[X]$, and, since $J(\ell_1\otimes X)$ contains all such sequences, it follows that J is an isometric isomorphism.

3.2 $C(K)$ and $\mathcal{L}_\infty$-spaces

We now investigate injective tensor products with the space $C(K)$ of continuous functions on a compact topological space K. We shall show that the injective tensor product $C(K)\hat{\otimes}_\varepsilon X$ can be identified with the Banach space $C(K,X)$ of continuous functions from K into X, where the norm on this space is given by $\|f\|_\infty=\sup\{\|f(t)\|:t\in K\}$. Consider the canonical linear mapping $J\colon C(K)\otimes X\to C(K,X)$ given by $Ju(t)=\sum_{i=1}^n f_i(t)x_i$, where $\sum_{i=1}^n f_i\otimes x_i$ is any representation of u. We show first that the norm of the function Ju coincides with the injective norm of the tensor u. We have

$$\|Ju\|_\infty=\sup_{t\in K}\left\|\sum_{i=1}^n f_i(t)x_i\right\|=\sup_{t\in K}\left\|\sum_{i=1}^n \delta_t(f_i)x_i\right\|,$$

where $\delta_t\in C(K)^*$ is the evaluation functional, $\delta_t(f)=f(t)$. Now the set $\{\delta_t:t\in K\}$ of all evaluation functionals clearly forms a norming set in $C(K)^*$, and so, using the identification of the injective norm with the operator norm in $\mathcal{L}(C(K)^*,X)$, we see that $\|Ju\|_\infty=\varepsilon(u)$.

It only remains to show that $J(C(K)\otimes X)$ is dense in $C(K,X)$. Let $f\in C(K,X)$ and let $\varepsilon>0$. Let $\{f(t_1),\ldots,f(t_n)\}$ be an ε-net for the compact set $f(K)$ and for each i, take $V_i=\{t\in K:\|f(t)-f(t_i)\|<\varepsilon\}$. Then the open sets $V_1,\ldots,V_n$ cover K. Let $\{g_1,\ldots,g_n\}$ be a partition of unity on K subordinate to this cover. Thus the $g_i\in C(K)$ take their values in the interval $[0,1]$, the support of g_i is contained in V_i and $\sum_{i=1}^n g_i(t)=1$ for every $t\in K$. Let $u=\sum_{i=1}^n g_i\otimes f(t_i)\in C(K)\otimes X$. Then $\|Ju-f\|=\|\sum_{i=1}^n g_i(t)f(t_i)-f(t)\|=\|\sum_{i=1}^n g_i(t)(f(t_i)-f(t))\|$. Now, fixing $t\in K$, then, for each i, either $\|f(t_i)-f(t)\|<\varepsilon$ or $g_i(t)=0$. Therefore

$\|Ju - f\| < \varepsilon$. It follows that J is an isometric isomorphism, and $C(K)\hat{\otimes}_\varepsilon X$ can be identified with $C(K, X)$.

This identification can be applied to obtain a representation of a space of continuous functions of two variables as an injective tensor product of two spaces of continuous functions. Let K_1 and K_2 be compact spaces. The space $C(K_1 \times K_2)$ can be canonically identified with the space $C(K_1, C(K_2))$; the function $f \in C(K_1 \times K_2)$ corresponds to the function f_1 defined by $f_1(t_1)(t_2) = f(t_1, t_2)$. Therefore we have

$$C(K_1)\hat{\otimes}_\varepsilon C(K_2) = C(K_1 \times K_2)\,.$$

We recall that projective tensor products with the space $L_1(\mu)$ respect subspaces. We propose to prove an analogous result for $C(K)$, namely that injective tensor products with $C(K)$ respect quotients. Our proof will shed some light on the structure of the space $C(K)$ itself. Before stating the result, we make the following simple observation. In order to show that an operator $R\colon W \to Z$ of unit norm is a quotient operator, it suffices to establish that, for every $z \in Z$ and every $\varepsilon > 0$, there exists $w \in W$ such that $\|Rw - z\| < \varepsilon$ and $\|w\| < (1 + \varepsilon)\|z\|$.

Proposition 3.5. *Let $Q\colon Z \to X$ be a quotient operator and let I be the identity operator on $C(K)$. Then $I \otimes_\varepsilon Q\colon C(K)\hat{\otimes}_\varepsilon Z \to C(K)\hat{\otimes}_\varepsilon X$ is a quotient operator.*

Proof. Since $C(K) \otimes X$ is dense in $C(K)\hat{\otimes}_\varepsilon X$, it suffices to prove that for every $u \in C(K) \otimes X$ and every $\delta > 0$, there exists $v \in C(K)\hat{\otimes}_\varepsilon Z$ such that $I \otimes Q(v) = u$ and $\varepsilon(v) \leq (1 + \delta)\varepsilon(u)$. Let $\{x_1, \ldots, x_n\}$ be a δ-net for the compact subset $Ju(K)$ of X, where Ju is the continuous function associated with u. For each i, let $V_i = \{t \in K : \|Ju(t) - x_i\| < \delta\}$, so that $\{V_1, \ldots, V_n\}$ is an open cover of K. Choose a partition of unity $\{g_1, \ldots, g_n\}$ subordinate to this cover, so that each g_i is a continuous function with values in $[0, 1]$ and support contained in V_i, and $\sum_{i=1}^n g_i(t) = 1$ for every $t \in K$. In addition, we may assume that there exist points $t_i \in V_i$ such that $g_i(t_i) = 1$ for each i. Consider the element $u' = \sum_{i=1}^n g_i \otimes x_i$ of $C(K)\hat{\otimes}_\varepsilon X$ and the associated function Ju'. We have $\varepsilon(u - u') < \delta$. Furthermore, $\|Ju'(t)\| \leq \max \|x_i\|$ and also $\|Ju'(t_i)\| = \|x_i\|$ for each i. It follows that $\varepsilon(u') = \|Ju'\|_\infty = \max \|x_i\|$. Now, since Q is a quotient operator, we may choose $z_1, \ldots, z_n \in Z$ such that $Qz_i = x_i$ and $\|z_i\| < (1+\delta)\|x_i\|$ for each i. Let $v = \sum_{i=1}^n g_i \otimes z_i \in C(K)\hat{\otimes}_\varepsilon Z$. Then $I \otimes_\varepsilon Q(v) = u'$ and $\varepsilon(v) = \|Jv\|_\infty = \max \|z_i\| < (1 + \delta)\varepsilon(u') \leq (1 + \delta)(\varepsilon(u) + \delta)$. Since $\delta > 0$ is arbitrary, this completes the proof. □

If we examine this proof to see why it works, we find that the crucial argument is a finite dimensional one. Indeed, the functions $g_1, \ldots, g_n$ of the partition of unity span a subspace of $C(K)$ that is a copy of the n-dimensional space ℓ_∞^n – the argument given in the proof shows that the norm in $C(K)$ of the linear combination $\sum_{i=1}^n a_i g_i$ is exactly $\|a\|_\infty = \max |a_i|$. The rest of the proof essentially takes place in the tensor product $\ell_\infty^n \otimes_\varepsilon X$.

Any space with the same finite dimensional structure as $C(K)$ will behave in the same way. We say that a Banach space X is a $\mathcal{L}_{\infty,\lambda}$*-space* for some $\lambda \geq 1$ if every finite dimensional subspace, M, of X is contained in a finite dimensional subspace, N, of dimension n, say, whose Banach–Mazur distance from ℓ_∞^n is at most λ. Thus, there exists an isomorphism $T: N \to \ell_\infty^n$ such that $\|T\|\|T^{-1}\| \leq \lambda$. The space X is called a $\mathcal{L}_\infty$*-space* if it is a $\mathcal{L}_{\infty,\lambda}$-space for some λ. Arguing as in Section 2.4, but with the approximation by simple functions replaced by partitions of unity, we find that the space $C(K)$ is a $\mathcal{L}_{\infty,1+\varepsilon}$-space for every $\varepsilon > 0$.

Let us agree on some terminology. We shall say that *injective tensor products with the space X respect quotients* if, for every quotient operator $Q: Z \to Y$, the tensor product operator $I \otimes_\varepsilon Q: X\hat{\otimes}_\varepsilon Z \to X\hat{\otimes}_\varepsilon Y$ is a quotient operator. And we say that injective tensor products with X respect quotients *isomorphically* if, for every quotient operator $Q: Z \to Y$, the tensor product operator $I \otimes_\varepsilon Q$ is surjective. This is equivalent to the statement that the space $X\hat{\otimes}_\varepsilon Y$ is isomorphic to the quotient space associated with the surjective operator $I \otimes_\varepsilon Q$.

Theorem 3.6.

(a) *If X is a $\mathcal{L}_\infty$-space, then injective tensor products with X respect quotients isomorphically.*

(b) *If X is a $\mathcal{L}_{\infty,1+\varepsilon}$-space for every $\varepsilon > 0$, then injective tensor products with X respect quotients.*

Proof. (a) Let $Q: Z \to Y$ be a quotient operator. Let $u \in X\hat{\otimes}_\varepsilon Y$ and let $\delta > 0$. Choose $u' \in X \otimes Y$ such that $\varepsilon(u - u') < \delta$. Then, fixing a representation of u', we obtain a finite dimensional subspace M of X, such that u' lies in $M \otimes_\varepsilon Y$, where its injective norm is the same. Now M is contained in a finite dimensional subspace, N, of X, of dimension n, say, for which there exists an isomorphism $T: M \to \ell_\infty^n$ with $\|T\|\|T^{-1}\| \leq \lambda$. We may assume, without loss of generality, that $\|T\| \leq 1$ and $\|T^{-1}\| \leq \lambda$. Consider the element $T \otimes I(u')$ of $\ell_\infty^n\hat{\otimes}_\varepsilon Y$. We have $\varepsilon(T \otimes I(u'); \ell_\infty^n \otimes Y) \leq \varepsilon(u')$. Since ℓ_∞^n is the space $C(K)$ where K is the discrete space with n points, it follows that $\ell_\infty^n\hat{\otimes}_\varepsilon Y$ is a quotient space of $\ell_\infty^n\hat{\otimes}_\varepsilon Z$ and so there exists $w \in \ell_\infty^n\hat{\otimes}_\varepsilon Z$ such that $I \otimes Q(w) = T \otimes I(u')$ and $\varepsilon(w) \leq (1+\delta)\varepsilon(u')$. Let $v = T^{-1} \otimes I(w) \in X\hat{\otimes}_\varepsilon Z$. Then $I \otimes Q(v) = u'$ and $\varepsilon(v) \leq \lambda(1+\delta)\varepsilon(u') \leq \lambda(1+\delta)(\varepsilon(u)+\delta)$. It follows that for every $\delta > 0$ there exists $z \in X\hat{\otimes}_\varepsilon Z$ such that $I \otimes Q(z) = u$ and $\varepsilon(z) \leq \lambda(1+\delta)\varepsilon(u)$.

(b) follows immediately from the proof of (a). □

3.3 $L_1(\mu)\hat{\otimes}_\varepsilon X$ and the Pettis Integral

Let (Ω, Σ, μ) be a measure space. We have already seen that the projective tensor product $L_1(\mu)\hat{\otimes}_\pi X$ can be identified with the space $L_1(\mu, X)$ of

Bochner integrable functions from Ω into X. We seek a similar representation for the injective tensor product $L_1(\mu)\hat{\otimes}_\varepsilon X$. This requires the development of a new integral.

We shall assume throughout this section that the measure μ is finite. We recall that a function $f\colon \Omega \to X$ is said to be weakly μ-measurable if the scalar valued function φf is μ-measurable for every $\varphi \in X^*$. The weakly μ-measurable function f is *weakly integrable* with respect to μ if the function φf is integrable for every $\varphi \in X^*$. We propose to show that we can define an integral for such functions, but that this integral lies in the bidual, X^{**}. To this end, consider the linear mapping $S\colon X^* \to L_1(\mu)$ given by $S\varphi = \varphi f$. To see that S is bounded, we apply the Closed Graph Theorem. Suppose that $f(\varphi_n)$ converges to φ in X^* and $(\varphi_n f)$ converges to $g \in L_1(\mu)$. The sequence $(\varphi_n f)$ has a subsequence $(\varphi_{n_k} f)$ that converges μ-almost everywhere to g. But $(\varphi_{n_k} f)$ converges pointwise to φf. Therefore $g = \varphi f$ and we have established that S is bounded. Now, since μ is finite, the dual space of $L_1(\mu)$ is $L_\infty(\mu)$ and so the adjoint operator $T = S^*\colon L_\infty(\mu) \to X^{**}$ satisfies

$$\langle \varphi, Tg\rangle = \int_\Omega g(\varphi f)\, d\mu$$

for $g \in L_\infty(\mu)$. It now follows, by taking $g = \chi_E$, that if f is weakly integrable then for every measurable subset E of Ω there is an element of X^{**}, which we denote by $\int_E f\, d\mu$, such that

$$\left\langle \varphi, \int_E f\, d\mu \right\rangle = \int_E \varphi f\, d\mu$$

for every $\varphi \in X^*$. The vector $\int_E f\, d\mu$ is called the *Dunford integral* of f over E.

To see that we do need to go to the bidual to define this integral, consider the function $f\colon [0,1] \to c_0$ defined by $f(t) = (\mu(E_n)^{-1}\chi_{E_n}(t))$, where μ is Lebesgue measure and (E_n) is any sequence of mutually disjoint Lebesgue measurable sets of positive Lebesgue measure whose union is $[0,1]$. It is easy to see that f is weakly μ-measurable, and if $\varphi = (\lambda_n) \in c_0^* = \ell_1$, then $\int_0^1 \varphi f\, d\mu = \sum_n \lambda_n$. Therefore the Dunford integral of f over $[0,1]$ is the element $(1, 1, \dots)$ of ℓ_∞.

We say that the weakly integrable function f is *Pettis integrable* if the integral $\int_E f\, d\mu$ belongs to X for every $E \in \Sigma$. In this case, the integral $\int_E f\, d\mu$ is referred to as the *Pettis integral* of f over E. When f is Pettis integrable, the operator T takes its values in X, since it is obvious that $Tg = \int_\Omega gf\, d\mu$ belongs to X for every measurable simple function, g, and these functions are dense in $L_\infty(\mu)$. The operator T is sometimes referred to as the *Dunford operator* of f. We can say a little more about this operator.

Proposition 3.7. *Let (Ω, Σ, μ) be a finite measure space and let $f\colon \Omega \to X$ be Pettis integrable with respect to μ. Then the operator $T\colon L_\infty(\mu) \to X$ defined by*

Now U is the composition of the operator $S\colon X^* \to L_1(\mu)$ given by $S\varphi = \varphi f$, with the operator from $L_1(\mu)$ into ℓ_1 that takes the function $g \in L_1(\mu)$ to $(\int_{B_n} g\,d\mu)$. By Proposition 3.7, S is weakly compact. Therefore U is also weakly compact and so, by the Schur property of ℓ_1, it follows that U is compact. Therefore the convergence of the series $\sum_{n=1}^{\infty} |\int_{B_n} \varphi f\,d\mu|$ is uniform in $\varphi \in B_{X^*}$ and so the series $\sum_{n=1}^{\infty} |\int_{B_n} \varphi_n f\,d\mu|$ must converge for every sequence $\{\varphi_n\}$ in B_{X^*}, a contradiction. □

If we now assume in addition that f is μ-measurable, we have:

Corollary 3.10. *Let (Ω, Σ, μ) be a finite measure space and let $f\colon \Omega \to X$ be a μ-measurable Pettis integrable function. Then there exists a sequence of μ-measurable simple functions that converges to f in the Pettis norm.*

Proof. Since f is μ-measurable, the sets $E_n = \{\omega \in \Omega : \|f(\omega)\| > n\}$ are measurable. Let $A_n = E_n \setminus E_{n+1}$, so that (A_n) is a disjoint sequence of measurable sets and $E_n = \bigcup_{k \geq n} A_k$ for every n. Then

$$|f\chi_{E_n}|_1 = \sup_{\varphi \in B_{X^*}} \int_{E_n} |\varphi f|\,d\mu \leq \sum_{k=n}^{\infty} |f\chi_{A_k}|_1\,,$$

and it follows from the proposition that $|f\chi_{E_n}|_1 \to 0$ as $n \to \infty$.

Now, given $\varepsilon > 0$, choose n so that $|f\chi_{E_n}|_1 < \varepsilon/2$. Then f is bounded on the complement of E_n and hence is Bochner integrable there. By the definition of the Bochner integral, there exists a μ-measurable simple function g, vanishing outside E_n, such that $\|f\chi_{E_n^c} - g\|_1 < \varepsilon/2$, where $\|\,.\,\|_1$ is the Bochner norm. Therefore

$$|f - g|_1 \leq |f\chi_{E_n^c} - g|_1 + |f\chi_{E_n}|_1 \leq \|f\chi_{E_n^c} - g\|_1 + |f\chi_{E_n}|_1 < \varepsilon\,.$$

□

We have seen that the Dunford operator $T\colon L_\infty(\mu) \to X$ of the Pettis integrable function f, given by $Tg = \int_\Omega gf\,d\mu$, is weakly compact. The behaviour of this operator improves along with that of f:

Proposition 3.11. *Let μ be a finite measure, let f be Pettis integrable with respect to μ and let $T\colon L_\infty(\mu) \to X$ be the weakly compact operator given by $Tg = \int_\Omega gf\,d\mu$.*

(a) *If f is μ-measurable, then T is compact.*
(b) *If f is Bochner integrable, then T is nuclear.*

Proof. (a) Let (f_n) be a sequence of μ-measurable simple functions that converges to f in the Pettis norm. Then the operators $T_n\colon L_\infty(\mu) \to X$ defined by $T_n g = \int_\Omega g f_n\,d\mu$ are of finite rank and $\|T - T_n\| = |f - f_n|_1 \to 0$.

Therefore T, being the limit in operator norm of a sequence of finite rank operators, is compact.

(b) Let $f \in L_1(\mu, X) = L_1(\mu)\hat{\otimes}_\pi X$. Then there are bounded sequences (f_n) and (x_n) in $L_1(\mu)$ and X respectively such that $\sum_n \|f_n\|_1\|x_n\| < \infty$ and $f = \sum_n f_n \otimes x_n$. Then Tg is given by the Bochner integral $\int_\Omega gf\, d\mu$ for $g \in L_\infty(\mu)$ and we have

$$\left\| \int_\Omega \sum_{k\geq n} gf_k x_k \, d\mu \right\| \leq \|g\|_\infty \sum_{k\geq n} \|f_k\|_1 \|x_k\| \to 0$$

as $n \to \infty$. Hence

$$Tg = \sum_{n=1}^{\infty} \left(\int_\Omega gf_n \, d\mu \right) x_n$$

for every $g \in L_\infty(\mu)$. Therefore T is nuclear. □

The fact that the Dunford operator of a μ-measurable Pettis integrable function is compact enables us to prove an important result that bridges the gap between weak and norm summability. A sequence (x_n) in a Banach space X is said to be *weakly subseries summable* if the series $\sum_n x_{k_n}$ is weakly convergent for every increasing sequence (k_n) of positive integers. Equivalently, the series $\sum_n \varepsilon_n x_n$ is weakly convergent for every choice of signs, $\varepsilon_n = \pm 1$.

Proposition 3.12 (The Orlicz–Pettis Theorem). *Every weakly subseries summable sequence in a Banach space is unconditionally summable in norm.*

Proof. Suppose there is a sequence (x_n) in the Banach space X that is weakly subseries summable but not unconditionally summable. Then, replacing (x_n) by a suitable sequence $(\varepsilon_n x_n)$ with $\varepsilon_n = \pm 1$, we may assume that (x_n) is weakly subseries summable, but that the series $\sum_n x_n$ diverges. Therefore there exist increasing sequences of positive integers, (m_k) and (n_k), with $m_k < n_k < m_{k+1}$ for every k, and $\delta > 0$, such that

$$\left\| \sum_{j=m_k}^{n_k} x_j \right\| \geq \delta$$

for every k. Let $z_k = \sum_{j=m_k}^{n_k} x_j$. Then the sequence (z_j) is weakly subseries summable, but $\|z_j\| \geq \delta$ for every j. A simple application of the Closed Graph Theorem shows that the weak sums of the form $\sum_j \varepsilon_j z_j$, where $\varepsilon_j = \pm 1$, lie in a bounded subset of X. Thus we can define a bounded function $f\colon [0,1] \to X$ by taking $f(t)$ to be the weak sum of the series $\sum_j r_j(t) z_j$, where $r_j(t)$ are the Rademacher functions. By the Pettis Measurability Theorem (Proposition 2.15), f is measurable with respect to Lebesgue measure. Therefore f

is Bochner integrable and so the Dunford operator $T\colon L_\infty[0,1] \to X$ is compact. But this forces the set $\{z_j\} = \{Tr_j\}$ to be relatively compact in X, which is impossible, since (z_j) converges weakly, but not in norm, to zero. □

We conclude with the representation of the injective tensor product $L_1(\mu)\hat{\otimes}_\varepsilon X$ as the completion of the space of μ-measurable Pettis integrable functions.

Proposition 3.13. *Let μ be a finite measure. The injective tensor product $L_1(\mu)\hat{\otimes}_\varepsilon X$ is isometrically isomorphic to the space $\hat{P}_1(\mu, X)$.*

Proof. Let $J\colon L_1(\mu) \otimes X \to P_1(\mu, X)$ be the canonical linear mapping and let $u = \sum_{i=1}^n f_i \otimes x_i \in L_1(\mu) \otimes X$. Then

$$\begin{aligned}
|Ju|_1 &= \sup_{\varphi \in B_{X^*}} \int_\Omega \left| \sum_{i=1}^n f_i \varphi(x_i) \right| d\mu = \sup_{\varphi \in B_{X^*}} \sup_{g \in B_{L_\infty(\mu)}} \left| \int_\Omega \sum_{i=1}^n g f_i \varphi(x_i)\, d\mu \right| \\
&= \sup_{\varphi \in B_{X^*}} \sup_{g \in B_{L_\infty(\mu)}} \left| \sum_{i=1}^n \langle g, f_i \rangle \langle \varphi, x_i \rangle \right| = \varepsilon(u)\,.
\end{aligned}$$

Therefore J is an isometry. Now the μ-measurable simple functions lie in the range of J and these functions are dense in $\hat{P}_1(\mu, X)$. Therefore J extends to an isometric isomorphism from $L_1(\mu)\hat{\otimes}_\varepsilon X$ into $\hat{P}_1(\mu, X)$. □

3.4 The Dual Space of $X\hat{\otimes}_\varepsilon Y$

We now turn our attention to the dual space of the injective tensor product $X\hat{\otimes}_\varepsilon Y$. Since the injective norm on $X \otimes Y$ is smaller than the projective norm, every bounded linear functional on $X\hat{\otimes}_\varepsilon Y$ is the linearization of a unique bounded bilinear form on $X \times Y$. We wish to characterize the bilinear forms that arise in this way.

Let us look again at the definition of the injective norm of an element $u = \sum_{j=1}^n x_j \otimes y_j$ of $X \otimes Y$:

$$\varepsilon(u) = \sup\left\{ \left| \sum_{i=1}^n \varphi(x_i)\psi(y_i) \right| : \varphi \in B_{X^*}, \psi \in B_{Y^*} \right\}.$$

This formula encourages us to think of u as a continuous function on the compact product space $B_{X^*} \times B_{Y^*}$, where B_{X^*} and B_{Y^*} carry their weak* topologies. Taking this point of view, we see that we have an embedding of $X \otimes_\varepsilon Y$ into the Banach space $C(B_{X^*} \times B_{Y^*})$, where this space carries the usual supremum norm. This embedding extends to the completion:

$$X\hat{\otimes}_\varepsilon Y \subset C(B_{X^*} \times B_{Y^*})\,.$$

This canonical embedding of the injective tensor product into a $C(K)$ space enables us to identify the dual. The Hahn–Banach Theorem allows us to consider a bounded linear functional on $X\hat{\otimes}_\varepsilon Y$ as a bounded linear functional on the space of continuous functions. Then the Riesz Representation Theorem shows that action of this functional is given by integration with respect to a suitable measure. To see how this works, suppose that B is a bounded bilinear form on $X \times Y$ whose linearization, $\tilde{B}$, is bounded for the injective norm on the tensor product. Then $\tilde{B}$ extends to a bounded linear functional on $C(B_{X^*} \times B_{Y^*})$ with the same norm. This functional is given by a regular Borel measure, μ, on $B_{X^*} \times B_{Y^{**}}$. Thus we have

$$\langle u, \tilde{B}\rangle = \int_{B_{X^*}\times B_{Y^*}} u(\varphi, \psi)\, d\mu(\varphi, \psi) \tag{3.5}$$

for every $u \in X\hat{\otimes}_\varepsilon Y$, and $\|\tilde{B}\| = \|\mu\|$, where $\|\mu\|$ is the variation norm of μ. In particular, if we take u to be an elementary tensor, we see that

$$B(x, y) = \int_{B_{X^*}\times B_{Y^*}} \varphi(x)\psi(y)\, d\mu(\varphi, \psi) \tag{3.6}$$

for every $x \in X$, $y \in Y$.

Conversely, if μ is a regular Borel measure on $B_{X^*} \times B_{Y^*}$, we may define a bounded linear functional on $X\hat{\otimes}_\varepsilon Y$ by (3.5) and the corresponding bilinear form B satisfies (3.6). Furthermore, a simple computation shows that $\|\tilde{B}\| \leq \|\mu\|$.

We have obtained a complete description of the dual space of $X\hat{\otimes}_\varepsilon Y$. Our findings are summarised in the following proposition.

Proposition 3.14. *Let B be a bilinear form on $X\times Y$. Then $\tilde{B}$ is a bounded linear functional on $X\hat{\otimes}_\varepsilon Y$ if and only if there exists a regular Borel measure μ on the compact space $B_{X^*} \times B_{Y^*}$ such that*

$$B(x, y) = \int_{B_{X^*}\times B_{Y^*}} \varphi(x)\psi(y)\, d\mu(\varphi, \psi)$$

for every $x \in X$, $y \in Y$. Furthermore, the norm of $\tilde{B}$ is given by

$$\|\tilde{B}\| = \inf \|\mu\|\,,$$

where μ ranges over the set of all measures that correspond to B in this way, and this infimum is attained.

We say that a bilinear form B on $X \times Y$ is an *integral bilinear form* if its linearization, $\tilde{B}$, is a bounded linear functional on the injective tensor product $X\hat{\otimes}_\varepsilon Y$. The *integral norm* of B is defined by

$$\|B\|_I = \inf \|\mu\|\,,$$

the infimum being taken over all the regular Borel measures on $B_{X^*} \times B_{Y^*}$ that satisfy (3.6). The Banach space of integral bilinear forms, with this norm, will be denoted by $\mathcal{B}_I(X \times Y)$. Thus, we have

$$(X\hat{\otimes}_\varepsilon Y)^* = \mathcal{B}_I(X \times Y).$$

We note some elementary properties of the class of integral bilinear forms. If B is an integral bilinear form on $X \times Y$ and $S\colon W \to X$ and $T\colon Z \to Y$ are operators, then the bilinear form B' defined on $W \times Z$ by $B'(w,z) = B(Sw, Tz)$, is integral and $\|B'\|_I \leq \|B\|_I \|S\| \|T\|$.

Next, we observe that every nuclear bilinear form is integral. Indeed, if B is nuclear, we may express it in the form

$$B(x,y) = \sum_{n=1}^{\infty} \lambda_n \varphi_n(x) \psi_n(y),$$

where (λ_n) is a summable sequence of scalars, and (φ_n), (ψ_n) are sequences of distinct unit vectors in X^*, Y^*, respectively. We may interpret this sum as an integral of the type given in (3.6), where μ is the measure with point masses λ_n at the points (φ_n, ψ_n). Therefore

$$\mathcal{B}_N(X \times Y) \subset \mathcal{B}_I(X \times Y).$$

Furthermore, we have $\|\mu\| = \sum_{n=1}^{\infty} |\lambda_n|$ and the sequence (λ_n) may be chosen so that $\sum_{n=1}^{\infty} |\lambda_n|$ is arbitrarily close to the nuclear norm of B. It follows from the definition of the integral norm that

$$\|B\|_I \leq \|B\|_N$$

for every integral bilinear form B.

Finally we recall that the injective tensor product respects subspaces; if X, Y are subspaces of W, Z respectively, then $X\hat{\otimes}_\varepsilon Y$ is a subspace of $W\hat{\otimes}_\varepsilon Z$. It follows by the Hahn–Banach Theorem that every integral bilinear form on $X \times Y$ can be extended to an integral bilinear form on $W \times Z$ with the same integral norm.

We now look at some examples.

Example 3.15. *Integral bilinear forms on $c_0 \times c_0$.*

We have seen that $c_0\hat{\otimes}_\varepsilon X = c_0(X)$. Therefore $(c_0\hat{\otimes}_\varepsilon c_0)^* = \ell_1(\ell_1) = \ell_1\hat{\otimes}_\pi \ell_1$ and so, in this case we have $\mathcal{B}_I(c_0 \times c_0) = \mathcal{B}_N(c_0 \times c_0)$ with equality of the nuclear and integral norms.

Example 3.16. *Integral bilinear forms on $C(K)$ spaces.*

Let K_1 and K_2 be compact spaces. We have seen that $C(K_1)\hat{\otimes}_\varepsilon C(K_2) = C(K_1 \times K_2)$ and it follows that a bilinear form B on $C(K_1) \times C(K_2)$ is integral if and only if there exists a necessarily unique regular Borel measure μ on $K_1 \times K_2$ such that

$$B(f,g) = \int_{K_1 \times K_2} f(s)g(t)\, d\mu(s,t)$$

for every $f \in C(K_1)$, $g \in C(K_2)$. By the uniqueness of the measure μ, we have $\|B\|_I = \|\mu\|$. A special case of this is the integral bilinear form on $C(K) \times C(K)$ given by

$$B(f,g) = \int_K f(t)g(t)\, d\mu(t)\,,$$

where μ is a regular Borel measure on K. Again, we have $\|B\|_I = \|\mu\|$.

This example is the prototype of all integral bilinear forms. It can viewed in purely measure theoretic terms:

Example 3.17. *The canonical integral bilinear form on $L_\infty(\mu) \times L_\infty(\mu)$.*

Let (Ω, Σ, μ) be a finite measure space. Consider the bilinear form $B\colon L_\infty(\mu) \times L_\infty(\mu) \to \mathbb{K}$ defined by

$$B(f,g) = \int_\Omega fg\, d\mu\,.$$

We shall show that B defines a bounded linear functional on the tensor product $L_\infty(\mu) \otimes_\varepsilon L_\infty(\mu)$. Since the measurable simple functions are dense in $L_\infty(\mu)$, it suffices to consider tensors of the form $u = \sum_{i=1}^n f_i \otimes g_i$, where f_i, g_i are measurable simple functions. Taking the canonical representations of f_i and g_i, we obtain a representation of u as a finite sum of the form $u = \sum_{j,k} c_{jk}\chi_{A_j} \otimes \chi_{B_k}$, where (A_j) and (B_k) are finite sequences of mutually disjoint measurable subsets of Ω, each having nonzero measure. Let us compute the injective norm of u. On the one hand, since the closed unit ball of $L_1(\mu)$ is a norming set in $L_\infty(\mu)^*$, we have

$$\begin{aligned}
\varepsilon(u) &= \sup\Big\{\Big|\sum_{j,k} c_{jk}\langle \chi_{A_j}, f\rangle\langle \chi_{B_k}, g\rangle\Big| : f,g \in L_1(\mu),\ \|f\|_1, \|g\|_1 \le 1\Big\} \\
&\le \sup\Big\{\sum_{j,k} |c_{jk}| \int_{A_j} |f|\, d\mu \int_{B_k} |g|\, d\mu : \|f\|_1, \|g\|_1 \le 1\Big\} \\
&\le \sup_{j,k} |c_{jk}| \sup\Big\{\Big(\sum_j \int_{A_j} |f|\, d\mu\Big)\Big(\sum_k \int_{B_k} |g|\, d\mu\Big) : \|f\|_1, \|g\|_1 \le 1\Big\} \\
&\le \sup_{j,k} |c_{jk}|\,.
\end{aligned}$$

Now fix j and k. Let $f = \mu(A_j)^{-1}\chi_{A_j}$ and $g = \mu(B_k)^{-1}\chi_{B_k}$. Then f and g are unit vectors in $L_1(\mu)$ and so

$$\varepsilon(u) \geq \left|\sum_{r,s} c_{rs}\langle \chi_{A_r}, f\rangle\langle \chi_{B_s}, g\rangle\right| = |c_{jk}|\,.$$

Therefore $\varepsilon(u) = \sup_{j,k}|c_{jk}|$. For such a tensor u, we have

$$\begin{aligned}|\langle u, B\rangle| &= \left|\sum_{j,k} c_{jk}\int_\Omega \chi_{A_j}\chi_{B_k}\,d\mu\right| = \left|\sum_{j,k} c_{jk}\mu(A_j\cap B_k)\right| \\ &\leq \sup_{j,k}|c_{jk}|\sum_{j,k}\mu(A_j\cap B_k) \leq \sup_{j,k}|c_{jk}|\mu(\Omega) = \varepsilon(u)\mu(\Omega)\,,\end{aligned}$$

and so B is bounded on $L_\infty(\mu)\otimes_\varepsilon L_\infty(\mu)$. This computation also shows that $\|B\|_I \leq \mu(\Omega)$. On the other hand, if we take $f = g = 1$, then the elementary tensor $u = f\otimes g$ has unit injective norm and hence $\langle u, B\rangle = \mu(\Omega) \leq \|B\|_I$. Therefore $\|B\|_I = \mu(\Omega)$.

Our next result shows that every integral bilinear form can be factored through the canonical integral bilinear form on a suitable $L_\infty(\mu)$.

Theorem 3.18. *Let X and Y be Banach spaces. A bilinear form B on $X\times Y$ is integral if and only if there exists a finite measure space (Ω, Σ, μ) and operators $S\colon X\to L_\infty(\mu)$, $R\colon Y\to L_\infty(\mu)$ such that*

$$B(x,y) = \int_\Omega (Sx)(Ry)\,d\mu$$

for every $x\in X$, $y\in Y$. Furthermore,

$$\|B\|_I = \inf\|S\|\|R\|\mu(\Omega)\,,$$

where the infimum is taken over all such factorizations of B, and this infimum is attained.

Proof. Suppose that B is integral. Then there exists a regular Borel measure, ν, on the compact space $B_{X^*}\times B_{Y^*}$ such that

$$B(x,y) = \int_{B_{X^*}\times B_{Y^*}} \varphi(x)\psi(y)\,d\nu$$

for every $x\in X$, $y\in Y$ and $\|B\|_I = \|\nu\|$. By the Radon–Nikodým Theorem, there is a Borel measurable function, h, on $B_{X^*}\times B_{Y^*}$ such that $|h(t)| = 1$ for every $t\in B_{X^*}\times B_{Y^*}$ and $d\nu = h\,d|\nu|$. Let $\mu = |\nu|$ and $\Omega = B_{X^*}\times B_{Y^*}$ with Σ the Borel σ-algebra. Let $S\colon X\to L_\infty(\mu)$, $R\colon Y\to L_\infty(\mu)$ be the operators given by $(Sx)(\varphi,\psi) = \varphi(x)$ and $(Ry)(\varphi,\psi) = h(\varphi,\psi)\psi(y)$. Then

$$B(x,y) = \int_\Omega (Sx)(Ry)\,d\mu$$

and, since $\|S\|, \|R\| \leq 1$, we have $\|B\|_I = \|\nu\| = \mu(\Omega) \geq \|S\|\|R\|\mu(\Omega)$.

Conversely, given such a factorization of the bilinear form B, it follows that B is integral and $\|B\|_I \leq \|S\|\|R\|\mu(\Omega)$. □

We note that, by multiplying R or S by a constant, it is possible to assume that the measure μ in this factorization is a probability measure.

We conclude with a look at some of the finite dimensional aspects of integral bilinear forms. If X and Y are finite dimensional, then $X \otimes_\varepsilon Y = \mathcal{B}(X^* \times Y^*)$ and thus $\mathcal{B}_I(X \times Y) = X^* \otimes_\pi Y^* = \mathcal{B}_N(X \times Y)$. Therefore, every bilinear form is integral and the integral and nuclear norms are the same. We shall see in Chapter 5 that this coincidence of the integral and nuclear bilinear forms is a phenomenon that can also occur in certain very special infinite dimensional spaces. However, in general, the spaces of integral and of nuclear bilinear forms are different.

We can find the influence of finite dimensional considerations at work in infinite dimensions. Indeed, a bilinear form B on the product $X \times Y$ of any two Banach spaces is integral if and only if there exists a positive constant C such that, for all finite dimensional subspaces M, N of X, Y respectively, we have

$$\|B_{M,N}\|_I \le C\,,$$

where $B_{M,N}$ denotes the restriction of B to $M \times N$. Furthermore, the integral norm of B is the infimum of all such C. This is an immediate consequence of the fact that the injective norm, like the projective norm, is finitely generated; in other words, we have

$$\varepsilon(u) = \inf\{\varepsilon_{M,N}(u) : M \subset X, N \subset Y\,,\ u \in M \otimes N\}$$

for every $u \in X \otimes Y$. The remarks above show that this property of the class of integral bilinear forms is not shared with the nuclear bilinear forms.

3.5 Integral Operators

We now examine the operators that correspond to the integral bilinear forms. Every operator $T\colon X \to Y$ corresponds to a bilinear form, $B_T\colon X \times Y^* \to \mathbb{K}$, defined by $B_T(x, \psi) = \langle Tx, \psi\rangle$. We say that T is an *integral operator* if the bilinear form B_T is integral and we define the *integral norm* of T, denoted by $\|T\|_I$, to be the integral norm of the bilinear form B_T. Clearly, we have $\|T\| \le \|T\|_I$. The space of integral operators from X into Y with this norm will be denoted by $\mathcal{I}(X, Y)$. The correspondence $T \mapsto B_T$ embeds $\mathcal{I}(X, Y)$ as a subspace of $\mathcal{B}_I(X \times Y^*)$. We leave it to the reader to verify that $\mathcal{I}(X, Y)$ is closed in $\mathcal{B}_I(X \times Y^*)$. Therefore $\mathcal{I}(X, Y)$ is a Banach space.

There is a plentiful supply of integral operators. Every nuclear operator is easily seen to be integral, with $\|T\|_I \le \|T\|_N$. Furthermore, if $T\colon X \to Y$ is integral and $V\colon W \to X$, $U\colon Y \to Z$ are operators, then the operator UTV is integral, with

$$\|UTV\|_I \le \|U\|\|T\|_I\|V\|\,.$$

If μ is a finite measure, then the canonical injection

$$I\colon L_\infty(\mu) \to L_1(\mu)$$

is integral, with integral norm equal to $\mu(\Omega)$, since the corresponding bilinear form is the canonical integral bilinear form on $L_\infty(\mu) \times L_\infty(\mu)$. This operator is the prototypical integral operator:

Theorem 3.19. *An operator $T\colon X \to Y$ is integral if and only if there exists a finite measure space (Ω, Σ, μ) and a pair of operators $S\colon X \to L_\infty(\mu)$, $R\colon L_1(\mu) \to Y^{**}$ such that $JT = RIS$, where J is the canonical embedding of Y into Y^{**} and I is the canonical mapping from $L_\infty(\mu)$ into $L_1(\mu)$. Furthermore,*

$$\|T\|_I = \inf \|S\|\|R\|\mu(\Omega)\,,$$

where the infimum is taken over all such factorizations of T, and this infimum is attained.

Proof. Suppose that T is integral. Applying Theorem 3.18 to the integral bilinear form B_T, there exists a finite measure space (Ω, Σ, μ) and operators $S\colon X \to L_\infty(\mu)$ and $U\colon Y^* \to L_\infty(\mu)$ such that

$$\langle Tx, \psi\rangle = \int_\Omega (Rx)(U\psi)\,d\mu$$

for every $x \in X$, $\psi \in Y^*$, with $\|T\|_I = \|B_T\|_I = \|S\|\|U\|\mu(\Omega)$. We obtain the required factorization by taking R to be the restriction of U^* to $L_1(\mu)$. Then, bearing in mind that the closed unit ball of $L_1(\mu)$ is a norming set in $L_\infty(\mu)^*$, we have $\|T\|_I = \|S\|\|R\|\mu(\Omega)$.

Conversely, suppose that there exists a factorization of the form $JT = RIS$. Let $V\colon Y^* \to L_\infty(\mu)$ be the restriction of R^* to Y^*. Then the bilinear form associated with T satisfies

$$\begin{aligned} B_T(x,\psi) &= \langle Tx, \psi\rangle = \langle RISx, \psi\rangle \\ &= \langle ISx, V\psi\rangle = \int_\Omega (Sx)(V\psi)\,d\mu\,. \end{aligned}$$

Therefore B_T is integral, and $\|B_T\|_I \le \|R\|\|V\|\mu(\Omega)$. □

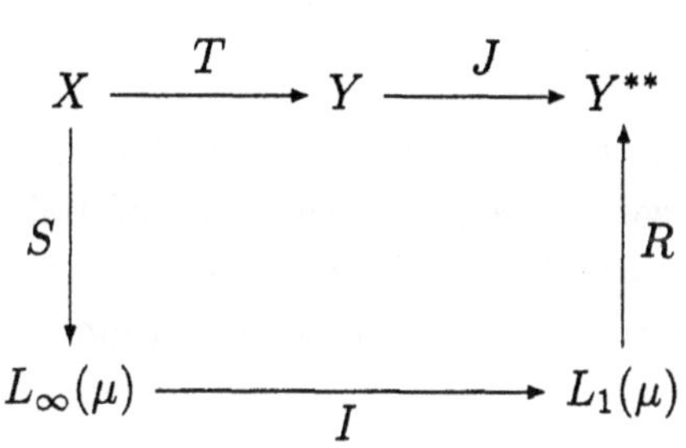

The factorization scheme given by the proposition is shown in the diagram above. An examination of the proof shows that the space $L_\infty(\mu)$ can be replaced by a $C(K)$ space, where the measure μ is now a positive regular Borel measure on K and I is the canonical mapping of $C(K)$ into $L_1(\mu)$, with the same statement regarding the integral norm of T. This factorization is shown below:

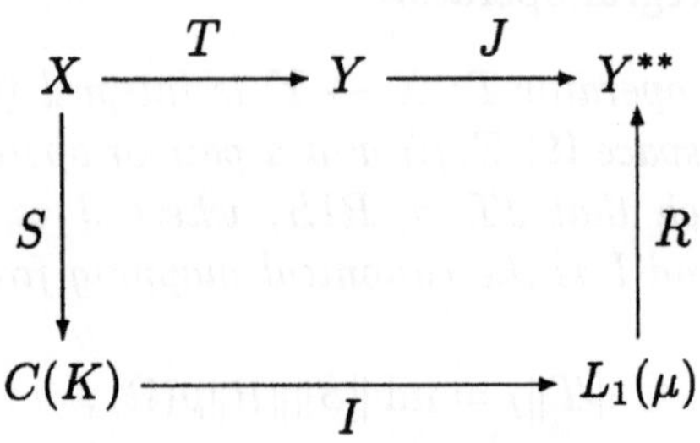

The next result illustrates the power of these factorizations. We recall that an operator is said to be *completely continuous* if it maps weakly convergent sequences into norm convergent sequences.

Proposition 3.20. *Every integral operator is both weakly compact and completely continuous.*

Proof. It suffices to prove these properties for either of the canonical operators $I\colon L_\infty(\mu) \to L_1(\mu)$ or $I\colon C(K) \to L_1(\mu)$. Now the operator $I\colon L_\infty(\mu) \to L_1(\mu)$ factors through the canonical injection $L_\infty(\mu) \to L_p(\mu)$ for every $p > 1$:

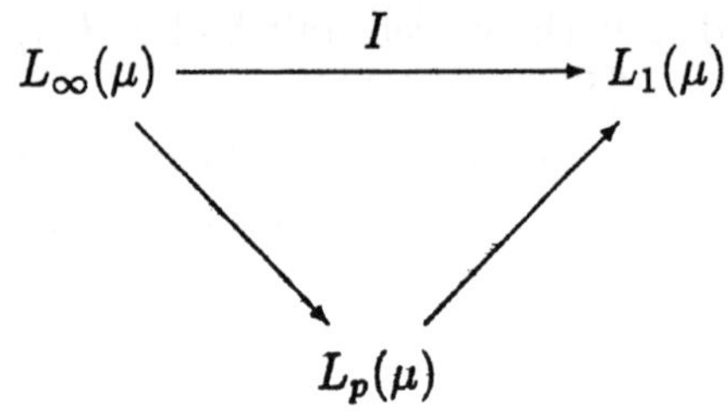

Hence, by the reflexivity of $L_p(\mu)$, the operator I is weakly compact.

To see that the canonical operator $I\colon C(K) \to L_1(\mu)$ is completely continuous, suppose that the sequence (f_n) converges weakly to zero in $C(K)$. Then the functions f_n are uniformly bounded and pointwise convergent to zero and it follows from the Bounded Convergence Theorem that (f_n) converges to zero in $L_1(\mu)$-norm. □

In the factorization scheme for an integral operator $T\colon X \to Y$, it is essential that we enlarge the range space to the bidual of Y. In some cases, it is possible to factor T directly. If Y is complemented in Y^{**} by a projection of norm one, we may compose JT with this projection to obtain a factorization of the form shown below:

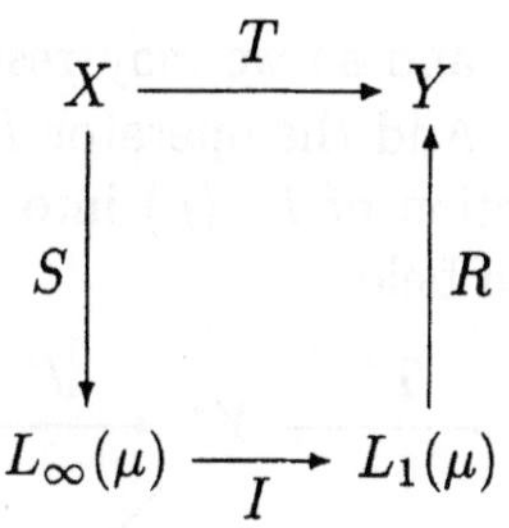

and we have $\|T\|_I = \inf \|R\|\|S\|\mu(\Omega)$, where the infimum is taken over all such factorizations. This applies in particular if Y is a dual space or an $L_1(\mu)$ space.

If we insist on factoring the operator T rather than JT, we obtain a smaller class of operators. We say that the operator $T\colon X \to Y$ is *Pietsch integral* if T can be factored in this way as RIS, and the *Pietsch integral norm* of T is defined to be $\|T\|_{PI} = \inf \|R\|\|S\|\mu(\Omega)$. The space of Pietsch integral operators from X into Y is denoted by $\mathcal{PI}(X,Y)$. This space is complete in the Pietsch integral norm. It is easy to see that

$$\mathcal{N}(X,Y) \subset \mathcal{PI}(X,Y) \subset \mathcal{I}(X,Y),$$

and $\|T\|_I \leq \|T\|_{PI} \leq \|T\|_N$ for every nuclear operator T. In general, these spaces of operators are distinct. From the remarks above, we see that the spaces $\mathcal{I}(X,Y)$ and $\mathcal{PI}(X,Y)$ are the same if Y is a dual space. Another special case occurs when the domain is c_0. We have seen that $c_0\hat{\otimes}_\varepsilon X = c_0(X)$ and hence

$$\mathcal{I}(c_0, X) = \mathcal{PI}(c_0, X) = \ell_1\hat{\otimes}_\pi X = \mathcal{N}(c_0, X),$$

with equality of the integral, Pietsch integral and nuclear norms.

Proposition 3.21. *Let $T\colon X \to Y$ be an operator and let $J\colon Y \to Y^{**}$ be the canonical inclusion. The following statements are equivalent:*

(i) *T is integral.*
(ii) *JT is integral.*
(iii) *T^* is integral.*
(iv) *T^{**} is integral.*

Furthermore, if T is integral, then

$$\|T\|_I = \|JT\|_I = \|T^*\|_I = \|T^{**}\|_I.$$

Proof. The equivalence of (i) and (ii), and the equality of the integral norms of T and JT, follow immediately from the remarks above concerning operators into a dual space.

(i) implies (iii): suppose that T is integral. Let $JT = RIS$, where $R\colon L_1(\mu) \to Y^{**}$ and $S\colon X \to L_\infty(\mu)$ are operators and $I\colon L_\infty(\mu) \to L_1(\mu)$ is the canonical injection. Then $T^*J^* = S^*I^*R^*$. Now J^* is the canonical projection of Y^{***} onto Y^* and so we may restrict R^* to obtain an operator U from Y^* into $L_\infty(\mu)$. And the operator $I^*\colon L_\infty(\mu) \to L_\infty(\mu)^*$ factors through the canonical injection of $L_\infty(\mu)$ into $L_1(\mu)$. Therefore, we have a factorization of T^* as shown below.

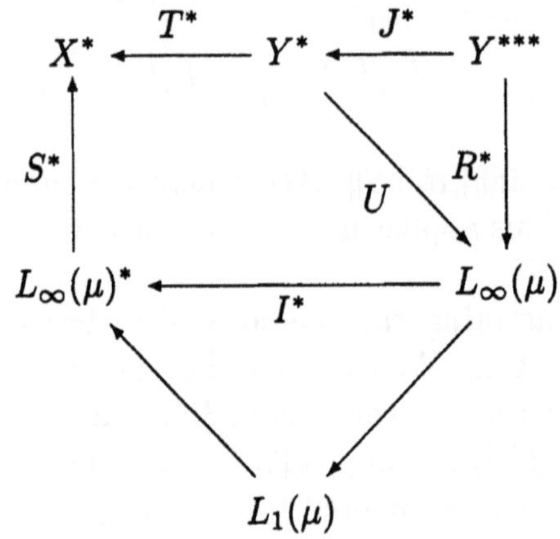

Hence T^* is integral and, since $\|U\| \le \|R^*\|$, it follows that $\|T^*\|_I \le \|T\|_I$. The proof that (i) implies (iii) shows that (iii) implies (iv) and that $\|T^{**}\|_I \le \|T^*\|_I$ whenever T^* is integral.

(iv) implies (i): Suppose that T^{**} is integral. Then there exists a factorization, $T^{**} = RIS$, where $R \in \mathcal{L}(L_1(\mu), Y^{**})$ and $S \in \mathcal{L}(X^{**}, L_\infty(\mu))$. Restricting T^{**} to X gives the operator $JT\colon X \to Y^{**}$. Hence we have a factorization of JT as shown below.

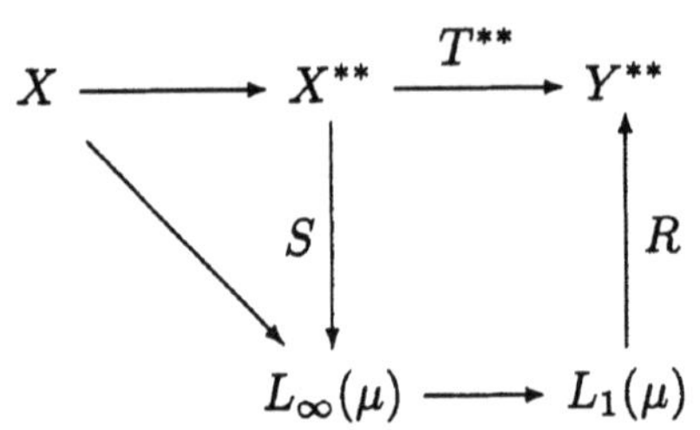

Therefore T is integral and $\|T\|_I \le \|T^{**}\|_I$.

We have shown that (i), (ii), (iii) and (iv) are equivalent and that, if T is integral, then

$$\|T\|_I \le \|T^{**}\|_I \le \|T^*\|_I \le \|T\|_I = \|JT\|_I .$$

Therefore these norms are equal. □

We recall that the space of bounded bilinear forms $\mathcal{B}(X \times Y)$ is isometrically isomorphic to the space $\mathcal{L}(X, Y^*)$. Under this identification, the bilinear form B is associated with the operator T, defined by $\langle Tx, y\rangle = B(x, y)$.

Proposition 3.22. *A bilinear form B on $X \times Y$ is integral if and only if the associated operator $T\colon X \to Y^*$ is integral. The mapping $B \mapsto T$ is an isometric isomorphism of $\mathcal{B}_I(X \times Y)$ with $\mathcal{I}(X, Y^*)$.*

Proof. Suppose that the bilinear form B is integral. Then there exist a probability space (Ω, Σ, μ), and operators $U\colon X \to L_\infty(\mu)$ and $V\colon Y \to L_\infty(\mu)$ such that $B(x,y) = \int_\Omega (Ux)(Vy)\,d\mu$ for every $x \in X$, $y \in Y$, and $\|B\|_I = \|U\|\|V\|$. Thus $\langle Tx, y\rangle = B(x,y) = \langle IUx, Vy\rangle = \langle V^*IUx, y\rangle$, where $I\colon L_\infty(\mu) \to L_1(\mu)$ is the canonical injection. Therefore $T = RIU$, where R denotes the restriction of V^* to the canonical image of $L_1(\mu)$ in $L_\infty(\mu)^*$, and it follows that T is integral, with $\|T\|_I \le \|B\|_I$.

Conversely, suppose that T is integral. Let $T = RIS$ be a factorization of T, where $R \in \mathcal{L}(L_\infty(\mu), Y^*)$ and $S \in \mathcal{L}(X, L_1(\mu))$, with $\|T\|_I = \|R\|\|S\|$. Then

$$\begin{aligned} B(x,y) &= \langle Tx, y\rangle = \langle RISx, y\rangle = \langle ISx, R^*y\rangle \\ &= \int_\Omega (Sx)(R^*y)\,d\mu \end{aligned}$$

for every $x \in X$, $y \in Y$. Therefore B is integral and $\|B\|_I \le \|S\|\|R^*\| = \|T\|_I$

It follows that B is integral if and only if T is integral and that $\|B\|_I$ is equal to $\|T\|_I$. □

It follows from this result that the dual space of the injective tensor product $X\hat{\otimes}_\varepsilon Y$ may be identified with the space $\mathcal{I}(X, Y^*)$:

$$(X\hat{\otimes}_\varepsilon Y)^* = \mathcal{B}_I(X \times Y) = \mathcal{I}(X, Y^*)\,.$$

In finite dimensions, every operator is integral and the integral norm coincides with the nuclear norm. Just as the integral bilinear forms can be characterized by their behaviour on finite dimensional subspaces, the integral operators are also "finitely determined". To make this precise, let $B\colon X \times Y^* \to \mathbb{K}$ be the bilinear form associated with the operator $T\colon X \to Y$. We have seen that B is integral if and only if there exists a positive constant, C, such that for every pair of finite dimensional subspaces, M, N of X, Y^* respectively, the restriction, $B_{M,N}$ of B to $M \times N$ satisfies $\|B_{M,N}\|_I \le C$, or equivalently, $\|B_{M,N}\|_N \le C$. Furthermore, the integral norm of B is the infimum of all such C. Now every finite dimensional subspace N of Y^* is the dual space of a quotient space X/F, where F is a subspace of X of finite codimension. Denoting the injection of a subspace G by I_G and the quotient operator with kernel G by Q_G, we have $Q_N = I_F^*$. Therefore the finite dimensional characterization of integral bilinear forms translates to operators in the following way:

Proposition 3.23. *An operator $T\colon X \to Y$ is integral if and only if there exists a positive constant C such that for every finite dimensional subspace E of X and every finite codimensional subspace F of Y, the finite dimensional operator*

$$Q_F T I_E\colon E \to X \to Y \to Y/F$$

satisfies $\|Q_F T I_E\|_N \le C$. Furthermore, we have $\|T\|_I = \inf C$, where the infimum is taken over all such pairs E, F.

This result suggests that the integral norm is the most natural extension to infinite dimensions of the nuclear norm for finite dimensional operators. We shall explore this idea further in a more general setting in due course.

3.6 Exercises

Exercise 3.1. Show that $c_0\hat{\otimes}_\varepsilon c_0$ is isometrically isomorphic to c_0.

Exercise 3.2. Show that $\ell_2\hat{\otimes}_\varepsilon\ell_2$ is not a Hilbert space.

Exercise 3.3. Let $Q\colon Z \to Y$ be a quotient operator and let I be the identity operator on X. Show that the tensor product operator $I \otimes_\varepsilon Q\colon X\hat{\otimes}_\varepsilon Z \to X\hat{\otimes}_\varepsilon Y$ is a quotient operator if the following "lifting" condition is satisfied: for every operator $S\colon X^* \to Y$ and every $\varepsilon > 0$, there exists an operator $T\colon X^* \to Z$ such that $QT = S$ and $\|T\| \le \|S\| + \varepsilon$.

Exercise 3.4. Let $Q\colon \ell_1 \to c_0$ be a quotient operator and let I be the identity operator on ℓ_1. Show that $I \otimes_\varepsilon Q\colon \ell_1\hat{\otimes}_\varepsilon\ell_1 \to c_0\hat{\otimes}_\varepsilon\ell_1$ is not a quotient operator.

Exercise 3.5. Identify the tensor diagonal in $\ell_p\hat{\otimes}_\varepsilon\ell_p$ for $1 \le p \le \infty$.

Exercise 3.6. (a) Show that if X is infinite dimensional then $\ell_\infty\hat{\otimes}_\varepsilon X \neq \ell_\infty(X)$.

(b) Find a description of $\ell_\infty\hat{\otimes}_\varepsilon X$ as a space of sequences.

Exercise 3.7. Show that $L_\infty(\mu)\hat{\otimes}_\varepsilon X$ is isometrically isomorphic to the space $K_\infty(\mu, X)$ of (equivalence classes of) μ-measurable essentially compact functions with values in X, where the norm of f is the essential supremum of $\|f\|$. (f is said to be essentially compact if there exists a μ-null set whose complement is mapped by f into a relatively compact subset of X.)

Exercise 3.8. Show that if X and Y have the Schur property, then so does $X\hat{\otimes}_\varepsilon Y$. (The Banach space X has the Schur property if every weakly convergent sequence in X converges in norm.)

Exercise 3.9. Let X and Y be $\mathcal{L}_\infty$-spaces. Show that $X\hat{\otimes}_\varepsilon Y$ is also a $\mathcal{L}_\infty$-space.

Exercise 3.10. Let K be a compact topological space and let $Q\colon Z \to X$ be a surjective operator. Show that every continuous function from K into X can be lifted to Z; in other words, if $f\colon K \to X$ is continuous, then there exists a continuous function $g\colon K \to Z$ such that $Qg = f$.

Exercise 3.11. Let $f\colon [0,1] \to L_1[0,1]$ be the function $f(t) = \chi_{[0,t]}$. Show that f is Pettis integrable with respect to Lebesgue measure on $[0,1]$ and compute the Pettis integral of f.

Exercise 3.12. Let $f = (f_n)\colon [0,1] \to c_0$, where f_n are Lebesgue integrable functions on $[0,1]$ with disjoint supports. Let μ be Lebesgue measure on $[0,1]$.

(a) Show that the function f is Pettis integrable with respect to μ if and only if $\sup \|f_n\|_1 < \infty$. If f is Pettis integrable, show that

$$|f|_1 = \sup_n \|f_n\|_1 .$$

(b) Show that f is Bochner integrable with respect to μ if and only if $\sum_n \|f_n\| < \infty$. If f is Bochner integrable, show that

$$\|f\|_1 = \sum_{n=1}^{\infty} \|f_n\|_1 .$$

Exercise 3.13. (a) Show that the bilinear form on $\ell_2 \times \ell_2$ given by $B(x,y) = \sum_n x_n y_n$ is not integral.

(b) Show that the same bilinear form is integral on $\ell_1 \times \ell_1$ and find its integral norm.

Exercise 3.14. Show that every nuclear operator T is Pietsch integral and that $\|T\|_{PI} \leq \|T\|_N$.

Exercise 3.15. Show that the space $\mathcal{PI}(X,Y)$ is complete in the Pietsch integral norm.

Exercise 3.16. Show that the canonical injection of ℓ_1 into c_0 is integral and find its integral norm.

4. The Approximation Property

In this chapter we introduce the approximation property for Banach spaces. The possession of this property leads to the resolution of several outstanding issues concerning projective and injective tensor products. We then consider the following question: when are the projective or injective tensor products of reflexive spaces themselves reflexive? A satisfactory answer requires the use of the approximation property. Finally, we study tensor products of Banach spaces with Schauder bases.

4.1 The Approximation Property

There are several unanswered questions in the previous chapters.

1. The space $\mathcal{N}(X,Y)$ of nuclear operators is a quotient of the projective tensor product $X^*\hat{\otimes}_\pi Y$. When do we have $\mathcal{N}(X,Y) = X^*\hat{\otimes}_\pi Y$?
2. If the operator T has a nuclear adjoint, is T itself nuclear?
3. The injective tensor product $X^*\hat{\otimes}_\varepsilon Y$ can be identified with the space of approximable operators from X into Y. Thus, $X^*\hat{\otimes}_\varepsilon Y$ is a subspace of the space $\mathcal{K}(X,Y)$ of compact operators. When do we have $\mathcal{K}(X,Y) = X^*\hat{\otimes}_\varepsilon Y$?
4. Every nuclear operator T is integral and $\|T\|_I \leq \|T\|_N$. When do the integral and nuclear norms coincide on $\mathcal{N}(X,Y)$?

These questions all reduce to one that is very simple to state. Recall that a finite rank tensor $u = \sum_{i=1}^n x_i \otimes y_i \in X \otimes Y$ is zero if and only if $\sum_{i=1}^n \varphi(x_i)y_i = 0$ for every $\varphi \in X^*$. Now consider the problem of identifying when an element of the completed tensor product $X\hat{\otimes}_\pi Y$ is zero. Let $\sum_{n=1}^\infty x_n \otimes y_n$ be a representation of the tensor $u \in X\hat{\otimes}_\pi Y$. Can we deduce from

$$\sum_{n=1}^{\infty} \varphi(x_n)y_n = 0 \qquad \text{for every } \varphi \in X^* \tag{4.1}$$

that $u = 0$? Since the dual space of $X\hat{\otimes}_\pi Y$ is $\mathcal{B}(X \times Y) = \mathcal{L}(X, Y^*)$, we have $u = 0$ if and only if

$$\sum_{n=1}^{\infty} \langle y_n, Tx_n \rangle = 0 \qquad \text{for every } T \in \mathcal{L}(X, Y^*) . \tag{4.2}$$

It follows easily from (4.1) that $\sum_{n=1}^{\infty} \langle y_n, Tx_n \rangle = 0$ for every finite rank operator T. Now, if X and Y are infinite dimensional, there will be operators that are not of finite rank. Thus, there is a gap between the conditions expressed in (4.1) and (4.2). This gap can be bridged if it is possible to approximate an arbitrary operator by an operator of finite rank, at least on the set $\{x_n\}$. Now, we may assume without loss of generality that the sequence (x_n) tends to zero. Then the set $\{x_n\}$ is relatively compact and so we arrive at the following condition on the space X: every operator from X into Y^* can be approximated on a compact subset of X by a finite rank operator. Of course, there is an equivalent formualtion of (4.1) that leads to a similar statement concerning operators from Y into X^*. Our first result shows that the approximability of operators either from X or into X on compact sets is equivalent to the approximability of the identity operator on X on compact sets.

Proposition 4.1. *The following are equivalent for a Banach space X:*

(i) *For every compact subset K of X and every $\varepsilon > 0$ there exists a finite rank operator $S \colon X \to X$ such that $\|x - Sx\| \le \varepsilon$ for every $x \in K$.*
(ii) *For every Banach space Y, every operator $T \colon X \to Y$, every compact subset K of X and every $\varepsilon > 0$, there exists a finite rank operator $S \colon X \to Y$ such that $\|Tx - Sx\| \le \varepsilon$ for every $x \in K$.*
(iii) *For every Banach space Y, every operator $T \colon Y \to X$, every compact subset K of Y and every $\varepsilon > 0$, there exists a finite rank operator $S \colon Y \to X$ such that $\|Ty - Sy\| \le \varepsilon$ for every $y \in K$.*

Proof. (i) implies (ii): let $T \in \mathcal{L}(X, Y)$, $T \neq 0$, let $K \subset X$ be compact and let $\varepsilon > 0$. By (a), there exists a finite rank operator $R \colon X \to X$ such that $\|x - Rx\| \le \varepsilon / \|T\|$ for every $x \in K$. Then $S = TR$ is a finite rank operator from X into Y and $\|Tx - Sx\| \le \varepsilon$ for every $x \in K$.

(i) implies (iii): let $T \in \mathcal{L}(Y, X)$, let $K \subset Y$ be compact and let $\varepsilon > 0$. Applying (i) to the compact subset $T(K)$ of X, there exists a finite rank operator $R \colon X \to X$ such that $\|x - Rx\| \le \varepsilon$ for every $x \in T(K)$. Then $S = RT$ is a finite rank operator from Y into X and $\|Ty - Sy\| \le \varepsilon$ for every $y \in K$.

Since each of (ii) and (iii) clearly implies (i), the proof is complete. □

A Banach space X is said to have the *approximation property* if the equivalent statements of this proposition hold for X. We shall see that this property is precisely what is needed to deduce (4.2) from (4.1). First, we look at some examples of spaces with the approximation property.

Example 4.2. *$C(K)$ has the approximation property.*

Let J be a compact subset of $C(K)$ and let $\varepsilon > 0$. By the Arzelà–Ascoli Theorem, J is equicontinuous and so there is a finite open cover $\{U_1, \ldots, U_n\}$ of K such that, for each i, if $s, t \in U_i$, then $|f(s) - f(t)| \leq \varepsilon$ for every $f \in J$. Let $\{g_1, \ldots, g_n\}$ be a partition of unity on K subordinate to the cover $\{U_1, \ldots, U_n\}$. Choose points $t_i \in U_i$ and let $S\colon C(K) \to C(K)$ be given by $Sf = \sum_{i=1}^n f(t_i)g_i$. Then T is a finite rank operator and $\|f - Tf\| \leq \varepsilon$ for every $f \in J$.

Before giving the next example, we give a sufficient condition for the approximation property that does not require any knowledge of the compact sets.

Proposition 4.3. *Suppose that there exists a uniformly bounded net (T_α) of finite rank operators on X such that $T_\alpha x \to x$ for every $x \in X$. Then X has the approximation property.*

Proof. Let (T_α) be a net of finite rank operators such that $\sup \|T_\alpha\| = C < \infty$ and $T_\alpha x \to x$ for every $x \in X$. Let K be a compact subset of X and let $\varepsilon > 0$. Choose a δ-net, $\{x_1, \ldots, x_n\}$ for K, where $\delta = \min\{\varepsilon/3, \varepsilon/3C\}$. There exists α_0 such that if $\alpha \geq \alpha_0$, then $\|x_i - T_\alpha x_i\| \leq \varepsilon/3$ for each i. Let $x \in K$ and choose i such that $\|x - x_i\| < \delta$. Then

$$\|x - T_{\alpha_0}x\| \leq \|x - x_i\| + \|x_i - T_{\alpha_0}x_i\| + \|T_{\alpha_0}x_i - T_{\alpha_0}x\| < \varepsilon\,.$$

□

We apply this to obtain some more examples of spaces with the approximation property.

Example 4.4. *Every Banach space with a Schauder basis has the approximation property.*

Let (e_n) be a Schauder basis for X and let e_n^* be the associated coordinate functionals. For each n, let P_n be the finite rank operator on X given by $P_n x = \sum_{i=1}^n e_i^*(x)x_i$. Then the operators P_n are uniformly bounded and $P_n x \to x$ for every $x \in X$. Therefore X has the approximation property.

It follows that the spaces c_0 and ℓ_p, for $1 \leq p < \infty$, all have the approximation property. Indeed, for every indexing set I, the spaces $c_0(I)$ and $\ell_p(I)$ $(1 \leq p < \infty)$ have the approximation property, since we may define a uniformly bounded net of finite rank operators, (P_F), indexed by the set of finite subsets of I, by $P_F x = \sum_{i \in F} e_i^*(x)e_i$ and we have $P_F x \to x$ for every x. In particular, we see that every Hilbert space has the approximation property.

We now show that the spaces $L_p(\mu)$, where μ is any measure, have the approximation property. In this case, we can use averaging operators to approximate the identity operator.

Example 4.5. *$L_p(\mu)$ has the approximation property for every $p \in [1, \infty)$.*

Let $1 \le p < \infty$. We construct a net of finite rank operators as follows. Let $A = \{A_1, \ldots, A_n\}$ be a finite collection of disjoint measurable sets of finite positive measure. Let T_A be the finite rank operator on $L_p(\mu)$ defined by

$$T_A f = \sum_{i=1}^{n} \mu(A_i)^{-1/p} \Big(\int_{A_i} f \, d\mu \Big) \chi_{A_i} .$$

Note that, since $\mu(A_i) < \infty$, the integral of f over A_i is defined. Indeed, we have $\int_{A_i} |f| \, d\mu \le (\int_{A_i} |f|^p \, d\mu)^{1/p}$. It also follows that $\|T_A\| = 1$ for every A. If A and B are two such collections of measurable sets, we define $A < B$ if every set in A is the union of a subcollection of B. This construction yields a directed set and we claim that for every $f \in L_p(\mu)$ the net $(T_A f)$ converges to f. In view of the uniform boundedness of the operators T_A, it suffices to prove this for the simple functions in $L_p(\mu)$. But every such function can be written as $f = \sum_{i=1}^{n} c_i \chi_{A_i}$ where $A = \{A_1, \ldots, A_n\}$ consists of disjoint measurable sets of finite positive measure and it is easy to see that $T_B f = f$ for $B > A$.

The spaces $C(K)^*$, $L_\infty(\mu)$ and $L_\infty(\mu)^*$ also have the approximation property. Thus, all the classical Banach spaces of sequences and functions encountered thus far, along with their duals, have the approximation property. There do exist spaces without this property; it is a deep result of Enflo that the space $\mathcal{L}(\ell_2, \ell_2)$ does not have the approximation property.

We now show that the approximation property meets our needs. In the proof of the next proposition we shall make use of the fact that every compact set in a Banach space is contained in the closed convex hull of a sequence that converges to zero.

Proposition 4.6. *Let X be a Banach space. The following are equivalent:*

(i) *X has the approximation property.*
(ii) *If $u = \sum_{n=1}^{\infty} \varphi_n \otimes x_n \in X^* \hat{\otimes}_\pi X$, where (φ_n), (x_n) are bounded sequences in X^*, X respectively with $\sum_{n=1}^{\infty} \|\varphi_n\| \|x_n\| < \infty$ and if $\sum_{n=1}^{\infty} \varphi_n(x) x_n = 0$ for every $x \in X$, then $u = 0$.*
(iii) *For every Banach space Y, if $u = \sum_{n=1}^{\infty} x_n \otimes y_n \in X \hat{\otimes}_\pi Y$, where (x_n), (y_n) are bounded sequences in X, Y respectively with $\sum_{n=1}^{\infty} \|x_n\| \|y_n\| < \infty$ and if $\sum_{n=1}^{\infty} \varphi(x_n) y_n = 0$ for every $\varphi \in X^*$, then $u = 0$.*
(iv) *For every Banach space Y, if $u = \sum_{n=1}^{\infty} x_n \otimes y_n \in X \hat{\otimes}_\pi Y$, where (x_n), (y_n) are bounded sequences in X, Y respectively with $\sum_{n=1}^{\infty} \|x_n\| \|y_n\| < \infty$ and if $\sum_{n=1}^{\infty} \psi(y_n) x_n = 0$ for every $\psi \in Y^*$, then $u = 0$.*

Proof. (i) implies (iv): suppose that X has the approximation property. Let $u = \sum_{n=1}^{\infty} x_n \otimes y_n \in X \hat{\otimes}_\pi Y$, where (x_n) and (y_n) are bounded sequences in X and Y such that $\sum_{n=1}^{\infty} \|x_n\| \|y_n\| < \infty$, and suppose that $\sum_{n=1}^{\infty} \psi(y_n) x_n = 0$ for every $\psi \in Y^*$. Assume, without loss of generality, that $x_n \to 0$ and $\sum_{n=1}^{\infty} \|y_n\| < \infty$. Let $T \in \mathcal{L}(X, Y^*) = (X \hat{\otimes}_\pi Y)^*$ and let $\varepsilon > 0$. Since X has the approximation property, there exists a finite rank operator, $S\colon X \to Y^*$, such that $\|Tx_n - Sx_n\| \le \varepsilon$ for every n. We have $Sx = \sum_{i=1}^m \varphi_i(x)\psi_i$, where $\varphi_i \in X^*$ and $\psi_i \in Y^*$. Now

$$\langle u, S\rangle = \sum_{n=1}^{\infty}\sum_{i=1}^{m} \varphi_i(x_n)\psi_i(y_n) = \sum_{i=1}^{m} \varphi_i\Big(\sum_{n=1}^{\infty} \psi_i(y_n)x_n\Big) = 0\,.$$

Therefore

$$|\langle u, T\rangle| \le |\langle u, T-S\rangle| + |\langle u, S\rangle| \le \sum_{n=1}^{\infty} \|Tx_n - Sx_n\|\|y_n\| \le \varepsilon \sum_{n=1}^{\infty} \|y_n\|\,.$$

Since ε is arbitrary, it follows that $\langle u, T\rangle = 0$ for every $T \in \mathcal{L}(X, Y^*)$. Therefore $u = 0$.

(iv) implies (iii): suppose that $u = \sum_{n=1}^{\infty} x_n \otimes y_n \in X\hat{\otimes}_\pi Y$ satisfies the conditions given in (c). Then, for every $\varphi \in X^*$ and $\psi \in Y^*$,

$$0 = \psi\Big(\sum_{n=1}^{\infty} \varphi(x_n)y_n\Big) = \sum_{n=1}^{\infty} \varphi(x_n)\psi(y_n) = \varphi\Big(\sum_{n=1}^{\infty} \psi(y_n)x_n\Big)\,.$$

Therefore $\sum_{n=1}^{\infty} \psi(y_n)x_n = 0$ for every $\psi \in Y^*$, and so, by (iv), we have $u = 0$.

(iii) implies (ii): if $u = \sum_{n=1}^{\infty} \varphi_n \otimes x_n$ satisfies $\sum_{n=1}^{\infty} \varphi_n(x)x_n = 0$ for every $x \in X$, then $\sum_{n=1}^{\infty} \theta(x_n)\varphi_n(x) = 0$ for every $x \in X$, $\theta \in X^*$. Therefore $\sum_{n=1}^{\infty} \theta(x_n)\varphi_n = 0$ in X^* for every $\theta \in X^*$ and it follows from (iii) that $\sum_{n=1}^{\infty} x_n \otimes \varphi_n = 0$ in $X\hat{\otimes}_\pi X^*$. Since the mapping $x \otimes \varphi \mapsto \varphi \otimes x$ is an isomorphism between $X\hat{\otimes}_\pi X^*$ and $X^*\hat{\otimes}_\pi X$, we have $u = 0$.

(ii) implies (i): suppose that X does not have the approximation property. Let E be the locally convex space of operators from X to itself, with the topology of uniform convergence on compact sets. Thus, the topology on E is generated by the continuous seminorms

$$p_K(T) = \sup\{\|Tx\| : x \in K\}\,,$$

where K ranges over the compact subsets of X. Let F be the subspace of E consisting of the finite rank operators. Then the statement that X does not have the approximation property is equivalent to the statement that the identity operator, I, does not belong to the closure of F. It follows from the Hahn–Banach Theorem that there exists a continuous linear functional Φ on E, such that $\Phi(T) = 0$ for every $T \in F$ and $\Phi(I) = 1$.

Since Φ is continuous, there exists a compact subset, K, of X such that $|\Phi(T)| \le p_K(T)$ for every $T \in E$. Since K is contained in the closed convex hull of a sequence (x_n) that converges to zero, we have

$$|\Phi(T)| \le \sup_n \|Tx_n\| \qquad \text{for every } T \in E\,.$$

We may consider the sequence (Tx_n) as an element of the Banach space $c_0(X)$. Let Z be the subset of $c_0(X)$ consisting of all such elements, as T ranges over E. Then Z is a subspace of $c_0(X)$ and we may view Φ as a bounded linear functional on Z. We use the Hahn–Banach Theorem again to extend this functional to $c_0(X)$. Therefore, since the dual space of $c_0(X)$ is $\ell_1(X^*)$, there exists $(\varphi_n) \in \ell_1(X^*)$ such that $\Phi(T) = \sum_{n=1}^{\infty} \varphi_n(Tx_n)$ for every $T \in E$. But then

$$\Phi(I) = \sum_{n=1}^{\infty} \varphi_n(x_n) = 1\,,$$

while $\Phi(T) = 0$ for every finite rank operator T implies that $\sum_{n=1}^{\infty} \varphi_n(x)x_n = 0$ for every $x \in X$. Therefore (ii) does not hold. □

An examination of this proof shows that, when applying one of the tests (ii), (iii) or (iv), it suffices to use a separating set of vectors. For example, if U is a separating subset of X^*, then the condition given in (iii) is equivalent to the requirement that $\sum_{n=1}^{\infty} \varphi(x_n)y_n = 0$ for every $\varphi \in U$.

We recall that $X^*\hat{\otimes}_\pi X$ is a subspace of $X^*\hat{\otimes}_\pi X^{**}$. Applying this fact, together with the observation that X is a separating subset of the dual of X^*, we obtain:

Corollary 4.7. *If X^* has the approximation property, then so does X.*

The converse of this result is false; there exist Banach spaces with the approximation property whose duals do not have this property. For example, the aforementioned example of Enflo, $\mathcal{L}(\ell_2, \ell_2)$, is the dual of the projective tensor product $\ell_2\hat{\otimes}_\pi\ell_2$, which has the approximation property (see Exercise 4.5).

We can now address the first of the questions posed at the beginning of this section.

Corollary 4.8. *Let X and Y be Banach spaces.*

(a) *If X^* or Y has the approximation property, then*

$$X^*\hat{\otimes}_\pi Y = \mathcal{N}(X, Y)\,.$$

(b) *If X^* or Y^* has the approximation property, then*

$$X^*\hat{\otimes}_\pi Y^* = \mathcal{B}_N(X \times Y)\,.$$

Proof. We prove (a), from which (b) follows easily. Recall that there is a quotient operator $J \colon X^*\hat{\otimes}_\pi Y \to \mathcal{N}(X, Y)$, which associates with the tensor $u = \sum_{n=1}^{\infty} \varphi_n \otimes y_n$ the nuclear operator $(Ju)(x) = \sum_{n=1}^{\infty} \varphi_n(x)y_n$. It follows from the proposition that if X^* or Y has the approximation property, then the kernel of J is trivial and hence J is an isometric isomorphism. □

The next result shows that the trace functional can be defined for nuclear operators on a Banach space X, provided X has the approximation property.

Corollary 4.9. *Let X be a Banach space with the approximation property. If $u = \sum_{n=1}^{\infty} \varphi_n \otimes x_n \in X^*\hat{\otimes}_\pi X$, where (φ_n), (x_n) are bounded sequences in X^*, X respectively with $\sum_{n=1}^{\infty} \|\varphi_n\|\|x_n\| < \infty$ and if $\sum_{n=1}^{\infty} \varphi_n(x)x_n = 0$ for every $x \in X$, then $\sum_{n=1}^{\infty} \varphi_n(x_n) = 0$.*

Proof. Let tr be the trace functional on $X^*\hat{\otimes}_\pi X$. Thus tr is the linearization of the bounded bilinear form on $X^* \times X$ given by $(\varphi, x) \mapsto \varphi(x)$. From the proposition, we have $u = 0$ and hence $\mathrm{tr}(u) = \sum_{n=1}^{\infty} \varphi_n(x_n) = 0$. □

We can now answer the second of our questions. If $T\colon X \to Y$ is a nuclear operator and $\sum_{n=1}^{\infty} \varphi_n \otimes y_n$ is a nuclear representation of T, then the adjoint operator is given by $T^*\psi = \sum_{n=1}^{\infty} \psi(y_n)\varphi_n$. It follows that T^* is nuclear, with $\|T^*\|_N \leq \|T\|_N$. To prove the converse, we need the approximation property.

Proposition 4.10. *Let X be a Banach space whose dual X^* has the approximation property. If the operator $T\colon X \to Y$ has a nuclear adjoint, then T is nuclear and $\|T\|_N = \|T^*\|_N$.*

Proof. We have $T^* \in \mathcal{N}(Y^*, X^*) = Y^{**}\hat{\otimes}_\pi X^*$, since X^* has the approximation property. We shall show that T^* lies in the closed subspace $Y\hat{\otimes}_\pi X^*$ of $Y^{**}\hat{\otimes}_\pi X^*$. By the Hahn–Banach Theorem, it suffices to show that every bounded linear functional on $Y^{**}\hat{\otimes}_\pi X^*$ that vanishes on $Y\hat{\otimes}_\pi X^*$ also vanishes on T^*. Let $S \in (Y^{**}\hat{\otimes}_\pi X^*)^* = \mathcal{L}(Y^{**}, X^{**})$ and suppose that S vanishes on $Y\hat{\otimes}_\pi X^*$, in other words, that $Sy = 0$ for every $y \in Y$. Now choose a nuclear representation $\sum_{n=1}^{\infty} y_n^{**} \otimes \varphi_n$ for T^*. The restriction of T^{**} to X is the operator T and so $\sum_{n=1}^{\infty} \varphi_n(x)y_n^{**}$ belongs to Y for every $x \in X$. Therefore

$$\sum_{n=1}^{\infty} \varphi_n(x)Sy_n^{**} = 0 \qquad \text{for every } x \in X.$$

It follows that the same statement holds if x is replaced by an arbitrary element x^{**} of X^{**}. Hence, using the approximation property of X^* and Proposition 4.6(iii), we get

$$\sum_{n=1}^{\infty} \varphi_n \otimes Sy^{**} = 0 \qquad \text{in } X^*\hat{\otimes}_\pi X^{**}.$$

Therefore, by Corollary 4.9,

$$\langle T^*, S\rangle = \sum_{n=1}^{\infty} \langle \varphi_n, Sy^{**}\rangle = 0\,.$$

Thus T^* belongs to $Y\hat{\otimes}_\pi X^*$ and so T is nuclear. In addition, we have $\|T\|_N = \|T^*\|_N$, since $Y\hat{\otimes}_\pi X^*$ is a subspace of $Y^{**}\hat{\otimes}_\pi X^*$. □

We come to our third question: when do we have $\mathcal{K}(X,Y) = X^*\hat{\otimes}_\varepsilon Y$? The problem now is to approximate an arbitrary compact operator in the operator norm by a finite rank operator. The approximation property allows us to approximate any operator by a finite rank operator on a compact set. We wish to move the compactness assumption from the domain of approximation to the operator. We shall show that there is an equivalent statement of the approximation property that has this form.

Let K be a convex, balanced, compact set in a Banach space X. The subspace of X spanned by K is the union of all the positive scalar multiples of K. This space is normed by the Minkowski functional of K:

$$\|x\|_K = \inf\{\lambda > 0 : x \in \lambda K\}$$

and the closed unit ball for this norm is K. The normed space defined in this way will be denoted by X_K.

Lemma 4.11. *Let K be a convex, balanced, compact subset of the Banach space X.*

(a) *The normed space X_K is complete.*
(b) *There exists a convex, balanced, compact subset L of X containing K such that K is compact in the Banach space X_L.*

Proof. (a) Suppose that X_K contains a Cauchy sequence (x_n) that does not converge. We may assume without loss of generality that $x_n \in K$ for every n. Since the injection $X_K \to X$ is bounded, (x_n) converges in the norm of X to some $x \in K$. Then the sequence $z_n = x_n - x$ is Cauchy in X_K but does not converge there, while $z_n \to 0$ in X. Now, choosing a subsequence, there exists $\delta > 0$ such that $\|z_n\|_K > \delta$ for every n. On the other hand, there is a positive integer N such that $\|z_n - z_m\|_K < \delta$ if $n, m \geq N$. Therefore $z_n - z_m \in \delta K$ for all $n, m \geq N$. Letting $m \to \infty$ we get $z_n \in \delta K$ for every $n \geq N$. Therefore $\|z_n\|_K \leq \delta$ for every $n \geq N$, a contradiction.

(b) Since K is compact, there exists a sequence (x_n) in X that converges to zero such that K lies in the closed convex hull of $\{x_n\}$. Choose an increasing sequence (λ_n) of positive scalars that diverges to infinity such that $\lambda_n x_n$ still converges to zero in X. Let L be the closed, convex, balanced hull of $\{\lambda_n x_n\}$. Then $\|x_n\|_L \leq \lambda_n^{-1}$ and so the sequence (x_n) converges to zero in X_L. Therefore K is compact in X_L. □

We can now give a formulation of the approximation property in terms of uniform approximation of compact operators. Recall that an operator is said to be approximable if it is the limit in the operator norm of a sequence of finite rank operators. We shall use the following property of adjoint operators in the proof: let $T\colon W \to Z$ be an injective operator. Then $T^*(Z^*)$ is dense in W^* for any locally convex topology with respect to which the dual space of W^* is W. In particular, this holds for the topology of uniform convergence on the compact subsets of W.

Proposition 4.12. *Let X be a Banach space.*

(a) *X has the approximation property if and only if for every Banach space Y, every compact operator from Y into X is approximable.*
(b) *X^* has the approximation property if and only if for every Banach space Y, every compact operator from X into Y is approximable.*

Proof. (a) Suppose that X has the approximation property. Let $T: Y \to X$ be a compact operator and let $\varepsilon > 0$. There exists a finite rank operator $R: X \to X$ such that $\|x - Rx\| < \varepsilon$ for every x in the relatively compact subset $T(B_Y)$ of X. Let $S = RT$. Then S has finite rank and $\|T - S\| \leq \varepsilon$.

Conversely, suppose that every compact operator with values in X is approximable. Let K be a compact subset of X and let $\varepsilon > 0$. By the lemma, there exists a convex, balanced, compact set L containing K such that K is compact in the Banach space X_L. Now, since the injection $I: X_L \to X$ is compact, there exists a finite rank operator $R: X_L \to X$ such that $\|I - R\| < \varepsilon/2$. Let $R = \sum_{i=1}^n \psi_i \otimes x_i$, where $\psi \in (X_L)^*$ and $x_i \in X$. Since $I: X_L \to X$ is injective, $I^*(X^*)$ is dense in $(X_L)^*$ for the topology of uniform convergence on the compact subsets of X_L. Therefore, given $\eta > 0$, there exist $\varphi_i \in X^*$ such that $\|\varphi_i(x) - \psi_i(x)\| < \eta$ for every $x \in K$ and every i. Let $S: X \to X$ be the finite rank operator $\sum_{i=1}^n \varphi_i \otimes x_i$. If $x \in K$, then

$$\|Rx - Sx\| \leq \sum_{i=1}^n |\varphi_i(x) - \psi_i(x)| \|x_i\| < \eta \sum_{i=1}^n \|x_i\| < \varepsilon\,,$$

taking $\eta < \varepsilon/(2\sum_{i=1}^n \|x_i\|)$. Since $\|x - Rx\| < \varepsilon/2$ for every $x \in K$, we have $\|x - Sx\| < \varepsilon$ for every $x \in K$. Therefore X has the approximation property.

(b) Suppose that X^* has the approximation property. Let $T: X \to Y$ be compact and let $\varepsilon > 0$. Since $T^*: Y^* \to X^*$ is compact, there exists a finite rank operator $R: X^* \to X^*$ such that $\|\varphi - R\varphi\| < \varepsilon$ for every φ in the relatively compact set $T^*(B_{Y^*})$. Since T is compact, T^{**} maps X^{**} into Y and so the composition $T^{**}R^*$ is a finite rank operator from X^{**} into Y. Let S be the restriction of this operator to X. We claim that this finite rank operator approximates T uniformly. If $x \in B_X$, then

$$\begin{aligned}\|Tx - Sx\| &= \sup_{\psi \in B_{Y^*}} |\psi Tx - \psi T^{**}R^*x| = \sup_{\psi \in B_{Y^*}} |(T^*\psi)x - (RT^*\psi)x| \\ &\leq \sup_{\psi \in B_{Y^*}} \|T^*\psi - R(T^*\psi)\| < \varepsilon\,.\end{aligned}$$

Therefore T is approximable.

Conversely, suppose that every compact operator with X as domain is approximable. We use (a) to show that X^* has the approximation property. Let $T: Y \to X^*$ be a compact operator. Let J_X denote the canonical embedding of X into X^{**}. By hypothesis, the compact operator $T^*J_X: X \to Y^*$ is approximable. It follows that the adjoint operator $(T^*J_X)^*: Y^{**} \to X^*$ is approximable. But T is the restriction of this operator to the canonical image of Y in Y^{**}. Therefore T is also approximable. Hence X^* has the approximation property. □

We now have the answer to our third question.

Corollary 4.13. *If either X^* or Y has the approximation property then $\mathcal{K}(X,Y) = X^*\hat{\otimes}_\varepsilon Y$.*

There is just one question remaining: under what conditions do the nuclear and integral norms coincide on the space $\mathcal{N}(X,Y)$ of nuclear operators? In order to give a satisfactory answer to this question, it is necessary to strengthen the approximation property. We say that the Banach space X has the *bounded approximation property* if there exists a positive constant λ such that, for every compact subset K of X and every $\varepsilon > 0$, there exists a finite rank operator $S\colon X \to X$ such that $\|S\| \leq \lambda$ and $\|x - Sx\| \leq \varepsilon$ for every $x \in K$. If this holds with $\lambda = 1$ then X is said to have the *metric approximation property*. A look back at the examples we have given shows that in every case the space in question has the metric approximation property. Indeed, the condition given in Proposition 4.3 actually implies the bounded approximation property and, moreover, if every operator in the net of approximating finite rank operators has norm at most one, then the space in question has the metric approximation property. Thus all the spaces $C(K)$, c_0, ℓ_p, $L_p(\mu)$, along with their duals, have the metric approximation property. If X has a Schauder basis (e_n), then the norm on X may be replaced by the equivalent norm

$$\|x\|' = \sup_n \|P_n(x)\|,$$

where P_n is the projection onto the first n coordinates, and X has the metric approximation property in this norm.

The equivalent formulations of the approximation property given in Proposition 4.1 also have metric versions. An inspection of the proof of that proposition shows that the metric approximation property for X is equivalent to the approximability of an operator T into or out of X on a compact set, by a finite rank operator S that satisfies $\|S\| \leq \|T\|$.

Theorem 4.14. *The following are equivalent for a Banach space X:*

(i) *X has the metric approximation property.*
(ii) *For every Banach space Y, the canonical mapping $X\hat{\otimes}_\pi Y \to (X^*\hat{\otimes}_\varepsilon Y^*)^*$ is an isometric embedding.*
(iii) *The canonical map $X\hat{\otimes}_\pi X^* \to (X^*\hat{\otimes}_\varepsilon X)^*$ is an isometric embedding.*

Proof. (i) implies (ii): the canonical mapping from $X\hat{\otimes}_\pi Y$ into $(X^*\hat{\otimes}_\varepsilon Y^*)^*$, or $\mathcal{B}_I(X^* \times Y^*)$, has norm one. Thus we must show that $\|u\|_I \geq \pi(u)$ for every $u \in X\hat{\otimes}_\pi Y$. Fix $u \in X\hat{\otimes}_\pi Y$ and $\varepsilon > 0$. Choose a representation $\sum_{n=1}^\infty x_n \otimes y_n$ of u with $x_n \to 0$ and $0 < \sum_{n=1}^\infty \|y_n\| < \infty$. Since the dual space of $X\hat{\otimes}_\pi Y$ is $\mathcal{L}(X, Y^*)$, there exists $T \in \mathcal{L}(X, Y^*)$ with $\|T\| = 1$, such that $|\langle u, T\rangle| \geq \pi(u) - \varepsilon$. Applying the metric approximation property, there exists a finite rank operator $S\colon X \to Y^*$ such that $\|S\| \leq 1$ and

$$\|Tx_n - Sx_n\| < \varepsilon\left(\sum_{n=1}^{\infty} \|y_n\|\right)^{-1}$$

for every n. Then $|\langle u, T - S\rangle| < \varepsilon$ and it follows that

$$|\langle u, S\rangle| \geq \pi(u) - 2\varepsilon\,.$$

Now $S\colon X \to Y^*$, being a finite rank operator, belongs to $X^*\hat{\otimes}_\varepsilon Y^*$, and since the operator norm coincides with the injective norm, we have $\varepsilon(S) \leq 1$. Therefore,

$$\|u\|_I \geq |\langle u, S\rangle| \geq \pi(u) - 2\varepsilon\,.$$

Since $\varepsilon > 0$ is arbitrary, it follows that $\|u\|_I = \pi(u)$.

(ii) implies (iii) is trivial.

(iii) implies (i): suppose that $X\hat{\otimes}_\pi X^*$ is a subspace of $(X^*\hat{\otimes}_\varepsilon X)^*$. Since the closed unit ball of $X^* \otimes_\varepsilon X$ is dense in the closed unit ball of the completed tensor product $X^*\hat{\otimes}_\varepsilon X$, it follows that

$$\pi(u) = \sup\{|\langle u, S\rangle| : S \in X^* \otimes X,\ \varepsilon(S) \leq 1\}$$

for every $u \in X\hat{\otimes}_\pi X^*$. On the other hand, we have

$$\pi(u) = \sup\{|\langle u, T\rangle| : T \in \mathcal{L}(X, X),\ \|T\| \leq 1\}\,.$$

Therefore, the set $F = \{S \in \mathcal{L}(X, X) : S$ has finite rank and $\|S\| \leq 1\}$ is weak*-dense in the closed unit ball of $\mathcal{L}(X, X)$, where these sets are taken as subsets of $(X\hat{\otimes}_\pi X^*)^*$. Thus, the identity operator, I, on X can be weak*-approximated by elements of F, so that if $x \in X$, $\varphi \in X^*$ and $\varepsilon > 0$, then there exists $S \in F$ such that $|\langle Sx, \varphi\rangle - \langle x, \varphi\rangle| < \varepsilon$. It follows that for every $x \in X$, the point x belongs to the weak closure of the set $Fx = \{Sx : S \in F\}$. But this set is convex, and so its weak and norm closures coincide. Thus, x lies in the norm closure of the set Fx for every x. Now consider the strong operator topology on $\mathcal{L}(X, X)$, generated by the seminorms $T \mapsto \|Tx\|$, as x ranges over X. We have shown that the identity belongs to the closure of F for this topology. Therefore there exists a net (S_α) of finite rank operators, each having norm at most one, such that $S_\alpha x \to x$ for every $x \in X$. Hence, by the remarks preceding the proposition, X has the metric approximation property. □

We recall from Proposition 3.21 that the integral norm of an integral operator $T\colon X \to Y$ is unchanged if we consider T as an integral operator from X into Y^{**}. Thus $\mathcal{I}(X, Y)$ is a subspace of $\mathcal{I}(X, Y^{**})$.

Corollary 4.15. *If X^* has the metric approximation property, so does X.*

Proof. Applying part (ii) of the proposition to X^*, with $Y = X$, we see that the canonical mapping of $X^*\hat{\otimes}_\pi X$ into $(X^{**}\hat{\otimes}_\varepsilon X^*)^* = \mathcal{I}(X^*, X^{***})$ is an isometric embedding. Now the elements of $X^*\hat{\otimes}_\pi X$, considered as integral operators from X^* into X^{***}, take their values in X^*. But $\mathcal{I}(X^*, X^*)$ is a subspace of $\mathcal{I}(X^*, X^{***})$. Therefore, the canonical mapping of $X^*\hat{\otimes}_\pi X$ into $\mathcal{I}(X^*, X^*) = (X\hat{\otimes}_\varepsilon X^*)^*$ is an isometric embedding. Hence, by statement (iii) of the proposition, X has the metric approximation property. □

When a dual space has the metric approximation property, we have another canonical embedding:

Proposition 4.16. *If the dual space X^* has the metric approximation property, then the canonical mapping $X^*\hat{\otimes}_\pi Y \to (X\hat{\otimes}_\varepsilon Y^*)^*$ is an isometric embedding for every Banach space Y.*

Proof. The canonical mapping $X^*\hat{\otimes}_\pi Y \to (X^{**}\hat{\otimes}_\varepsilon Y^*)^* = \mathcal{I}(Y^*, X^{***})$ is an isometric embedding. But the elements of $X^*\hat{\otimes}_\pi Y$, considered as integral operators from Y^* into X^{***}, take their values in X^*. Therefore $X^*\hat{\otimes}_\pi Y$ is isometrically embedded into the subspace $\mathcal{I}(Y^*, X^*) = (X\hat{\otimes}_\varepsilon Y^*)^*$ of $(X^{**}\hat{\otimes}_\varepsilon Y^*)^*$ □

We can now answer our last question.

Corollary 4.17. *If either X^* or Y has the metric approximation property, then $\mathcal{N}(X, Y)$ is a closed subspace of $\mathcal{I}(X, Y)$.*

Proof. Since X^* or Y has the approximation property, $\mathcal{N}(X, Y) = X^*\hat{\otimes}_\pi Y$. We have seen that, if either X^* or Y has the metric approximation property, then the canonical mapping of $X^*\hat{\otimes}_\pi Y$ into $(X\hat{\otimes}_\varepsilon Y^*)^* = \mathcal{I}(X, Y^{**})$ is an isometric embedding. Arguing as in the proof of the proposition, it follows that the canonical mapping of $\mathcal{N}(X, Y)$ into $\mathcal{I}(X, Y)$ is an isometric embedding. □

There are spaces with the approximation property that fail to have the metric approximation property. However, in the next chapter we shall prove a striking result of Grothendieck to the effect that every reflexive space, and every separable dual space, with the approximation property has the metric approximation property.

4.2 Reflexivity of Tensor Products

We have seen that the projective tensor product of reflexive spaces can fail to be reflexive – the projective tensor product of two Hilbert spaces, for example, contains a copy of ℓ_1 and so cannot be reflexive. For the same reason, $L_{p_1}[0,1]\hat{\otimes}_\pi L_{p_2}[0,1]$ contains a complemented isomorphic copy of ℓ_1 for every p_1, p_2 and so cannot be reflexive. In this section, we find sufficient conditions for the tensor products $X\hat{\otimes}_\pi Y$ and $X\hat{\otimes}_\varepsilon Y$ to be reflexive and we give some examples of pairs of spaces that satisfy these conditions.

Since the dual space of $X\hat{\otimes}_\pi Y$ is $\mathcal{L}(X, Y^*)$, the reflexivity of $X\hat{\otimes}_\pi Y$ is equivalent to that of $\mathcal{L}(X, Y^*)$. We shall begin with a sufficient condition for the space of operators to be reflexive. First, we need a characterization of weak convergence in the space of compact operators.

Lemma 4.18. *Let X be a reflexive Banach space. A bounded sequence (T_n) in $\mathcal{K}(X, Y)$ converges weakly to $T \in \mathcal{K}(X, Y)$ if and only if the sequence $(T_n x)$ converges weakly to Tx for every $x \in X$.*

Proof. This condition is clearly necessary. To establish sufficiency, consider the compact space $K = B_X \times B_{Y^*}$, where B_X and B_{Y^*} carry the weak and weak* topologies respectively. Each compact operator $T\colon X \to Y$ defines a continuous function f_T on K, given by $f_T(x, \psi) = \langle Tx, \psi\rangle$, and $\|T\| = \|f_T\|_\infty$. Therefore $\mathcal{K}(X, Y)$ may be regarded as a closed subspace of $C(K)$. Now let (T_n) be a bounded sequence in $\mathcal{K}(X, Y)$ and suppose that $T_n x$ converges weakly to Tx for every $x \in X$. Then the sequence of functions f_{T_n} converges pointwise to the function f_T. Furthermore, this sequence is bounded in $C(K)$. It follows from the Riesz Representation Theorem for the dual space $C(K)^*$ and the Bounded Convergence Theorem that f_{T_n} converges weakly to f_T in $C(K)$. Therefore (T_n) converges weakly to T in $\mathcal{K}(X, Y)$. □

Theorem 4.19. *Let X and Y be reflexive Banach spaces. If every operator from X into Y is compact, then $\mathcal{L}(X, Y)$ is reflexive.*

Proof. Let (T_n) be a sequence in the closed unit ball of $\mathcal{L}(X, Y)$. We shall prove that (T_n) has a weakly convergent subsequence.

Suppose to begin with that X is separable. Let $\{x_k\}$ be a countable, dense subset of X. By the reflexivity of Y, the sequence $(T_n x_k)_n$ has a weakly convergent subsequence for every k. Hence, using a diagonal argument, there exists a subsequence (T_{n_j}) of (T_n) such that the sequence $(T_{n_j} x_k)_j$ converges for every k. It follows that the sequence $(T_{n_j} x)$ converges weakly for every $x \in X$. Indeed, let $x \in X$ and let $\varepsilon > 0$. Choosing k such that $\|x - x_k\| < \varepsilon/2$, we have

$$T_{n_r} x - T_{n_s} x = (T_{n_r} - T_{n_s})(x - x_k) + (T_{n_r} x_k - T_{n_s} x_k)$$

for every r and s. Since

$$\|(T_{n_r} - T_{n_s})(x - x_k)\| < \varepsilon$$

for every r, s, it follows that the sequence $(T_{n_j} x)$ is weakly Cauchy in Y.

We may now define $T\colon X \to Y$ by taking Tx to be the weak limit of the sequence $(T_{n_j} x)$ and it follows from the Uniform Boundedness Principle that $T \in \mathcal{L}(X, Y) = \mathcal{K}(X, Y)$. Applying the lemma, we see that (T_{n_j}) converges weakly to T.

We now drop the condition that X be separable. Each of the adjoint operators $T_n^*\colon Y^* \to X^*$ is compact and as a result, has a separable range. Therefore, there exists a separable subspace Z of X^*, such that $T_n^*(Y^*) \subset Z$ for every n. Let Q be the quotient operator from X onto $X/Z^\perp$. Now $X/Z^\perp$ is a separable reflexive space, since its dual space is Z. Furthermore, each of the operators T_n factors through this quotient space. To see this, let $x \in X$ and suppose that $Qx = 0$, that is, $x \in Z^\perp$. Then, for every $\psi \in Y^*$, we have

$$\langle T_n x, \psi\rangle = \langle x, T_n^* \psi\rangle = 0\,.$$

It follows that $T_n x = 0$ for every n. Therefore, there exist operators $S_n\colon X/Z^\perp \to Y$ such that $S_n Q = T_n$ and $\|S_n\| \le 1$ for every n. Now $Q(B_X)$ is dense in the closed unit ball of $X/Z^\perp$ and so every operator from $X/Z^\perp$ into Y is compact. By the first part of the proof, (S_n) has a subsequence (S_{n_j}) that converges weakly to some operator S. Then, by the lemma, the sequence (T_{n_j}) converges weakly to $T = SQ$. □

The converse holds if we assume that one of the spaces has the approximation property.

Theorem 4.20. *Let X and Y be reflexive Banach spaces, one of which has the approximation property. If $\mathcal{L}(X,Y)$ is reflexive, then every operator from X to Y is compact.*

Proof. Suppose there is an operator $T\colon X \to Y$ that is not compact. Then there is a sequence (x_n) in the unit ball of X that converges weakly to some x, for which the sequence (Tx_n) does not converge in norm. Taking a suitable subsequence of the sequence $(x_n - x)$, we obtain a sequence (w_n) that converges weakly to zero and $\delta > 0$, such that $\|Tw_n\| \ge \delta$ for every n. Choose $\psi_n \in S_{Y^*}$ such that $\psi_n(Tw_n) = \|Tw_n\|$ for every n. Now $(X\hat{\otimes}_\pi Y^*)^* = \mathcal{L}(X,Y)$ and so $X\hat{\otimes}_\pi Y^*$ is reflexive. Therefore, the bounded sequence $(w_n \otimes \psi_n)$ has a subsequence, $(w_{n_k} \otimes \psi_{n_k})$, that converges weakly to some $u \in X\hat{\otimes}_\pi Y^*$. Since $\langle w_{n_k} \otimes \psi_{n_k}, T\rangle \ge \delta$ for every k, it follows that $\langle u, T\rangle \ge \delta$ also.

If $\varphi \in X^*$ and $y \in Y$, then

$$\langle w_{n_k} \otimes \psi_{n_k}, \varphi \otimes y\rangle = \varphi(w_{n_k})\psi_{n_k}(y) \to 0\,,$$

and hence $\langle u, \varphi \otimes y\rangle = 0$. Thus, u is zero when considered as a bilinear form on $X^* \times Y$. However, since X or Y^* has the approximation property, the canonical mapping of $X\hat{\otimes}_\pi Y^*$ into $\mathcal{B}(X^* \times Y)$ is injective. Therefore $u = 0$ in $X\hat{\otimes}_\pi Y^*$, which implies $\langle u, T\rangle = 0$, a contradiction. □

We now translate these results back into the language of tensor products. We shall make use of a theorem of Grothendieck, to be proved in the next chapter (Theorem 5.50), to the effect that every reflexive Banach space with the approximation property has the metric approximation property.

Theorem 4.21. *Let X and Y be reflexive Banach spaces, one of which has the approximation property. Then the following are equivalent:*

(i) $X\hat{\otimes}_\pi Y$ *is reflexive.*
(ii) $(X\hat{\otimes}_\pi Y)^* = X^*\hat{\otimes}_\varepsilon Y^*$.
(iii) *Every operator from X to Y^* is compact.*
(iv) $X^*\hat{\otimes}_\varepsilon Y^*$ *is reflexive.*

Proof. (i) implies (ii): if $X\hat{\otimes}_\pi Y$ is reflexive, then so is $(X\hat{\otimes}_\pi Y)^* = \mathcal{L}(X, Y^*)$. Therefore

$$(X\hat{\otimes}_\pi Y)^* = \mathcal{L}(X, Y^*) = \mathcal{K}(X, Y^*) = X^*\hat{\otimes}_\varepsilon Y^*,$$

since either X^* or Y^* has the approximation property.

(ii) implies (iii) is trivial.

(iii) implies (iv): if every operator from X into Y^* is compact, then $X^*\hat{\otimes}_\varepsilon Y^* = \mathcal{K}(X, Y^*) = \mathcal{L}(X, Y^*)$ is reflexive by Theorem 4.19.

(iv) implies (i): If $X^*\hat{\otimes}_\varepsilon Y^*$ is reflexive, then, by Proposition 4.16, the subspace $X\hat{\otimes}_\pi Y$ of the reflexive space $(X^*\hat{\otimes}_\varepsilon Y^*)^*$ is also reflexive. □

We have seen that the projective tensor product $L_{p_1}[0,1]\hat{\otimes}_\pi L_{p_2}[0,1]$ is never reflexive. It follows immediately from this proposition that the same is true for the injective tensor product:

Corollary 4.22. *The injective tensor product $L_{p_1}[0,1]\hat{\otimes}_\varepsilon L_{p_2}[0,1]$ is not reflexive for any p_1, p_2.*

After all these negative results, it is a pleasant suprise to find some non-trivial examples of reflexive tensor products.

Theorem 4.23 (Pitt's theorem). *If $1 < p < q < \infty$ then every operator from ℓ_q into ℓ_p is compact.*

Proof. Suppose that there is an operator $T\colon \ell_q \to \ell_p$ that is not compact. We may assume that $\|T\| = 1$. Then, arguing as in the proof of Theorem 4.20, there is a weakly null sequence (x_n) in ℓ_q and $\delta > 0$, such that $\delta < \|x_n\| \leq 1$ and $\|Tx_n\| > \delta$ for every n. For each m, let P_m and Q_m be the projections on either ℓ_p or ℓ_q defined by

$$P_m x = \sum_{i=1}^{m} x_i e_i \quad \text{and} \quad Q_m x = \sum_{i>m} x_i e_i .$$

We begin by choosing a positive integer s_1 such that

$$\|Q_{s_1} x_1\|_q < \delta/2 .$$

Let $z_1 = P_{s_1} x_1$. Then $\|z_1\|_q > \delta/2$ and $\|Tz_1\|_p > \delta/2$.

Since (x_n) converges weakly to zero, there is a positive integer $n_2 > 1$ such that

$$\|P_{s_1}x_{n_2}\|_q < \delta/4\,.$$

Choose $s_2 > s_1$ such that $\|Q_{s_2}x_{n_2}\|_q < \delta/4$ and let

$$z_2 = x_{n_2} - P_{s_1}x_{n_2} - Q_{s_2}x_{n_2}\,.$$

Then

$$\|z_2\|_q > \|x_{n_2}\|_q - \|P_{s_1}x_{n_2}\|_q - \|Q_{s_2}x_{n_2}\|_q > \delta/2\,.$$

Since $\|Tx_{n_2}\|_p > \delta$ and $\|T\| = 1$, we also have $\|Tz_2\|_p > \delta/2$.

Proceeding in this way, we obtain a sequence (z_k) in ℓ_q and an increasing sequence (s_k) of positive integers, with the following properties: $\delta/2 < \|z_k\|_q \leq 1$ and $\|Tz_k\|_p > \delta/2$, and the support of z_k lies in the range $s_{k-1} + 1 \leq i \leq s_k$.

We may now apply a similar technique to the sequence (Tz_k) in ℓ_p, since this sequence also converges weakly to zero. Thus, we obtain an increasing sequence (r_m) of positive integers and a subsequence (z_{k_m}) of (z_k), such that

$$\|P_{r_{m-1}}Tz_{k_m}\|_p < \delta/2^{m+2} \quad \text{and} \quad \|Q_{r_m}Tz_{k_m}\|_p < \delta/2^{m+2}$$

for every m. To simplify our notation, let us write $y_m = z_{k_m}$ and $w_m = Tz_{k_m} - u_m$, where

$$u_m = P_{r_{m-1}}Tz_{k_m} + Q_{r_m}Tz_{k_m}\,.$$

We now have

$$Ty_m = w_m + u_m\,,$$

where the vectors w_m are disjointly supported and

$$\|w_m\|_p \geq \|Ty_m\|_p - \|u_m\|_p > \delta/4$$

for every m. Furthermore, we have

$$\sum_{m=1}^{\infty} \|u_m\| \leq \sum_{m=1}^{\infty} \frac{\delta}{2^{m+1}} < \frac{\delta}{2}\,.$$

Now let λ be any element of the closed unit ball of ℓ_q that is not in ℓ_p. Let $a = \sum_{m=1}^{\infty} \lambda_m y_m$. Since the vectors y_m are disjointly supported, we have

$$\sum_{j=1}^{\infty} |a_j|^q = \sum_{m=1}^{\infty} |\lambda_m|^q \|y_m\|_q^q \leq \sum_{m=1}^{\infty} |\lambda_m|^q\,,$$

and so the series $\sum_{m=1}^{\infty} \lambda_m y_m$ converges to a in ℓ_q. Therefore the series $\sum_{m=1}^{\infty} \lambda_m Ty_m$ converges to Ta in ℓ_p. However,

$$\begin{aligned}
\left\|\sum_{m=1}^{M} \lambda_m T y_m\right\|_p &\geq \left\|\sum_{m=1}^{M} \lambda_m w_m\right\|_p - \left\|\sum_{m=1}^{M} \lambda_m u_m\right\|_p \\
&= \left(\sum_{m=1}^{M} |\lambda_m|^p \|w_m\|_p^p\right)^{1/p} - \left\|\sum_{m=1}^{M} \lambda_m u_m\right\|_p \\
&\geq \frac{\delta}{4}\left(\sum_{m=1}^{M} |\lambda_m|_p\right)^{1/p} - \sum_{m=1}^{M} \|u_m\|_p \\
&\geq \frac{\delta}{4}\left(\sum_{m=1}^{M} |\lambda_m|_p\right)^{1/p} - \frac{\delta}{2}
\end{aligned}$$

for every M. Hence $\|\sum_{m=1}^{M} \lambda_m T y_m\|_p \to \infty$ as $M \to \infty$, a contradiction. □

If $p \leq q$, then there is an injection of ℓ_p into ℓ_q that is not compact. This gives a complete description of the reflexive tensor products of ℓ_p spaces.

Corollary 4.24. *Let $1 < p, q < \infty$, with conjugate indices p', q'.*

(a) *$\ell_p \hat{\otimes}_\pi \ell_q$ is reflexive if and only if $p > q'$ and if this holds, then*

$$(\ell_p \hat{\otimes}_\pi \ell_q)^* = \ell_{p'} \hat{\otimes}_\varepsilon \ell_{q'} .$$

(b) *$\ell_p \hat{\otimes}_\varepsilon \ell_q$ is reflexive if and only if $p' > q$ and if this holds, then*

$$(\ell_p \hat{\otimes}_\varepsilon \ell_q)^* = \ell_{p'} \hat{\otimes}_\pi \ell_{q'} .$$

4.3 Tensor Product Bases

In this section we shall consider the formation of tensor products of Schauder bases. We begin by recalling some elementary facts. The sequence (e_n) is a Schauder basis for the Banach space X if every $x \in X$ is the sum of a convergent series of the form $\sum_{n=1}^{\infty} x_n e_n$ and the scalars x_n are uniquely determined by x. The coordinate functionals $e_n^* \in X^*$ are defined by $e_n^*(x) = x_n$ and two uniformly bounded sequences of projections, P_n and Q_n, are defined by $P_n x = \sum_{i=1}^{n} x_i e_i$ and $Q_n x = x - P_n x = \sum_{n+1=1}^{\infty} x_i e_i$. The number $\sup_n \|P_n\|$ is called the basis constant and the basis is said to be monotone if the basis constant is one. The formula $\|x\|' = \sup_n \|P_n x\|$ defines an equivalent norm on X relative to which the basis (e_n) is monotone.

The following criterion will be useful: a sequence (e_n) in a Banach space X is a Schauder basis for X if and only if $\{e_n\}$ spans a dense subspace of X and there exists a positive constant C, such that

$$\left\|\sum_{i=1}^{m} a_i e_i\right\| \leq C \left\|\sum_{i=1}^{n} a_i e_i\right\|$$

for every sequence of scalars (a_i) and every m, n with $m < n$.

Now let X, Y be Banach spaces with Schauder bases (e_n), (f_n) respectively. The tensor products $e_n \otimes f_m$ can be ordered as shown in the following diagram.

$$\begin{array}{ccccccc}
e_1 \otimes f_1 & \rightarrow & e_1 \otimes f_2 & & e_1 \otimes f_3 & & e_1 \otimes f_4 \\
 & & \downarrow & & \downarrow & & \downarrow \\
e_2 \otimes f_1 & \leftarrow & e_2 \otimes f_2 & & e_2 \otimes f_3 & & e_2 \otimes f_4 \\
 & & & & \downarrow & & \downarrow \\
e_3 \otimes f_1 & \leftarrow & e_3 \otimes f_2 & \leftarrow & e_3 \otimes f_3 & & e_3 \otimes f_4 \\
 & & & & & & \downarrow \\
 & & & \ldots & e_4 \otimes f_3 & \leftarrow & e_4 \otimes f_4
\end{array}$$

Thus, the ordering is $e_1 \otimes f_1$, $e_1 \otimes f_2$, $e_2 \otimes f_2$, $e_2 \otimes f_1$, $e_1 \otimes f_3$, $e_2 \otimes f_3$, ... This ordering is known as the *square ordering.*

We shall denote by P_n^1 and P_n^2 the projections onto the first n coordinates in X and Y respectively. Thus, $P_n^1 x = \sum_{i=1}^n e_i^*(x) e_i$ and $P_n^2 y = \sum_{i=1}^n f_i^*(y) f_i$. If $u = \sum_{i,j} a_{ij} e_i \otimes f_j$ is a finite sum in $X \otimes Y$, then the coefficients a_{ij} are given by $a_{ij} = e_i^* \otimes f_j^*(u)$. Therefore, the projection of u onto the first k coordinates is given by

$$P_k u = \sum_{(i,j) \in I_k} e_i^* \otimes f_j^*(u)\,,$$

where I_k is the set consisting of the first k ordered pairs of indices (i, j) in the square ordering. Now a glance at the diagram illustrating this ordering shows that the projection P_k can take one of three forms:

$$\begin{array}{rl}
P_n^1 \otimes P_n^2\,, & \text{if } k = n^2\,, \\
P_n^1 \otimes P_n^2 + P_{k-n^2}^1 \otimes (f_{n+1}^* \otimes f_{n+1})\,, & \text{if } n^2 < k \le n^2 + n + 1\,, \\
P_{n+1}^1 \otimes P_{n+1}^2 - (e_{n+1}^* \otimes e_{n+1}) \otimes P_{(n+1)^2-k}^2\,, & \text{if } n^2 + n + 1 < k < (n+1)^2.
\end{array}$$

Let A and B be the basis constants of the bases for X and Y respectively. Then $\|e_n^* \otimes e_n\| \le 2A$ and $\|f_n^* \otimes f_n\| \le 2B$. Now both the injective and projective norms satisfy $\|S \otimes T\| = \|S\|\|T\|$ where S and T are operators. Therefore $\|P_k\| \le 3AB$ in both cases. Now, it is easy to see that the tensor products $e_n \otimes f_m$ span dense subspaces of both the projective and injective tensor products. Thus, we have

Proposition 4.25. *Let X, Y be Banach spaces with Schauder bases (e_n), (f_n) respectively. Then the sequence $(e_n \otimes f_m)$, with the square ordering, is a Schauder basis for both $X \hat{\otimes}_\pi Y$ and $X \hat{\otimes}_\varepsilon Y$.*

We shall refer to the Schauder basis $(e_n \otimes f_m)$, with the square ordering, as the *tensor product basis.*

Recall that the basis (e_n) for X is said to be shrinking if the sequence (e_n^*) of coordinate functionals form a Schauder basis for the dual space X^*. This is equivalent to the condition that, for every $\varphi \in X^*$, the norm of the restriction of φ to the subspace $Q_n^1 X = (I - P_n^1)X$ of X tends to zero as $n \to \infty$. Or, to put it another way, if (x_n) is a bounded sequence in X such that $P_n^1 x_n = 0$ for every n, then x_n tends weakly to zero.

There is a simple test for weak convergence in an injective tensor product. The idea is much the same as that used in the proof of Lemma 4.18. Recall that $X\hat{\otimes}_\varepsilon Y$ may be regarded as a subspace of $C(K)$, where K is the product $B_{X^*} \times B_{Y^*}$ of the weak*-compact dual unit balls. Weak convergence of a bounded sequence in a $C(K)$ space is equivalent to pointwise convergence. Therefore, a bounded sequence (u_n) in $X\hat{\otimes}_\varepsilon Y$ converges weakly to zero if and only if $u_n(\varphi, \psi) \to 0$ for every $\varphi \in X^*$, $\psi \in Y^*$. We may consider the elements of $X\hat{\otimes}_\varepsilon Y$ either as compact operators from X^* into Y, or as compact operators from Y^* into X. This enables us to give two equivalent formulations of the criterion for weak convergence:

Proposition 4.26. *Let X, Y be Banach spaces. The following are equivalent for a bounded sequence (u_n) in $X\hat{\otimes}_\varepsilon Y$:*

(i) *(u_n) converges weakly to zero in $X\hat{\otimes}_\varepsilon Y$.*
(ii) *For every $\varphi \in X^*$, the sequence $(u_n\varphi)$ converges weakly to zero in Y.*
(iii) *For every $\psi \in Y^*$, the sequence $(u_n\psi)$ converges weakly to zero in X.*

We can now show that injective tensor products preserve the shrinking property.

Proposition 4.27. *Let X and Y be Banach spaces with shrinking Schauder bases (e_n) and (f_n) respectively. Then the tensor product basis for $X\hat{\otimes}_\varepsilon Y$ is shrinking.*

Proof. Let (u_n) be a bounded sequence in $X\hat{\otimes}_\varepsilon Y$ with the property that $P_n u_n = 0$ for every n, where P_n are the projections onto the first n coordinates for the tensor product basis $(e_i \otimes f_j)$. For each n, let $k(n)$ be the largest positive integer such that $k(n)^2 \le n$. Then $P_{k(n)}^1 \otimes P_{k(n)}^2(u_n) = 0$ and so

$$u_n = I \otimes (I - P_{k(n)}^2)(u_n) + (I - P_{k(n)}^1) \otimes P_{k(n)}^2(u_n) = u_n' + u_n''.$$

Now, for every $\varphi \in X^*$, the sequence $(u_n'\varphi)$ in Y is bounded and has the property that $P_k^2(u_n'\varphi) = 0$ if $n \ge k^2$. It follows from the shrinking property of the basis (f_j) that $(u_n'\varphi)$ tends weakly to zero. Therefore (u_n') converges weakly to zero in $X\hat{\otimes}_\varepsilon Y$. Similarly, for every $\psi \in Y^*$, the sequence $(u_n''\psi)$ tends weakly to zero in X and so the sequence (u_n'') converges weakly to zero in $X\hat{\otimes}_\varepsilon Y$. Therefore $u_n = u_n' + u_n''$ converges weakly to zero. □

While the injective tensor product preserves the shrinking property of bases, the projective tensor product preserves the dual property of bounded completeness. Recall that the Schauder basis (e_n) for the Banach space X is boundedly complete if X is the dual of a space with a shrinking basis and (e_n) is the dual basis. Equivalently, X is the dual of the closed subspace of X^* spanned by the coordinate functionals e_n^*.

Proposition 4.28. *Let X and Y be Banach spaces with boundedly complete Schauder bases (e_n) and (f_n) respectively. Then the tensor product basis for $X\hat{\otimes}_\pi Y$ is boundedly complete.*

Proof. There are Banach spaces W, Z, with shrinking bases (u_n), (v_n), respectively, such that $W^* = X$, $Z^* = Y$, and $u_n^* = e_n$, $v_n^* = f_n$ for every n. Replacing the norm of X by an equivalent norm, we may assume that the basis (e_n) is monotone. Then W^* has the metric approximation property, and so $X\hat{\otimes}_\pi Y = W^*\hat{\otimes}_\pi Z^*$ is a closed subspace of $(W\hat{\otimes}_\varepsilon Z)^*$. But $(e_n \otimes f_m) = (u_n^* \otimes v_m^*)$ is a Schauder basis both for $X\hat{\otimes}_\pi Y$ and for $(W\hat{\otimes}_\varepsilon Z)^*$. Therefore $X\hat{\otimes}_\pi Y = (W\hat{\otimes}_\varepsilon Z)^*$. □

The proof of this proposition gives some more information about the nuclear and integral bilinear forms on spaces with shrinking bases:

Corollary 4.29. *Let X and Y be Banach spaces with shrinking Schauder bases, at least one of which is monotone. Then every integral bilinear form on $X \times Y$ is nuclear and the integral and nuclear norms coincide.*

We conclude with a negative result. A Schauder basis (e_n) for a Banach space X is unconditional if, for every $x \in X$, the expansion of x relative to this basis is unconditionally convergent. The following example shows that the tensor product of unconditional bases need not be unconditional, even for the most well behaved spaces.

Example 4.30. *The standard tensor product bases for $\ell_2\hat{\otimes}_\pi\ell_2$ and $\ell_2\hat{\otimes}_\varepsilon\ell_2$ are not unconditional.*

Let (e_n) be the standard orthonormal basis for ℓ_2. We partition the set of positive integers into consecutive blocks of length 2^n: let $I_1 = \{1,2\}$, $I_2 = \{3,4,5,6\}$, and so on. Let

$$x = \sum_{n=1}^{\infty} \frac{1}{n} u_n\,, \quad \text{where } u_n = 2^{-n/2} \sum_{j \in I_n} e_j\,.$$

Since (u_n) is an orthonormal sequence, this series defines an element of ℓ_2. Now consider the expansion of the tensor $x \otimes x \in \ell_2\hat{\otimes}_\pi\ell_2$ relative to the tensor product basis. We have

$$x \otimes x = \sum_{n,m} x_n x_m e_n \otimes e_m\,.$$

This series converges for the square ordering. We claim that the convergence is not unconditional. We shall establish this by constucting a bounded bilinear form, B, on $\ell_2 \times \ell_2$ for which the series $\sum x_n x_m B(e_n, e_m)$ is not absolutely convergent.

Consider the bilinear form on the 2-dimensional Hilbert space ℓ_2^2 given by the matrix

$$A_1 = \begin{pmatrix} 1 & 1 \\ -1 & 1 \end{pmatrix}.$$

We have $\|A_1\| = 2^{1/2}$. For $n > 1$, let A_n be the $2^n \times 2^n$ matrix defined recursively by

$$A_n = \begin{pmatrix} A_{n-1} & A_{n-1} \\ -A_{n-1} & A_{n-1} \end{pmatrix}.$$

Then A_n defines a bilinear form on the 2^n-dimensional Hilbert space $\ell_2^{2^n}$ with norm $2^{n/2}$. Now let B be the bilinear form on ℓ_2 given by taking the direct sum of the unit norm bilinear forms $2^{-n/2}A_n$. Thus, B is given by the block diagonal matrix

$$\begin{pmatrix} 2^{-1/2}A_1 & & & & \\ & 2^{-1}A_2 & & & \\ & & \ddots & & \\ & & & 2^{-n/2}A_n & \\ & & & & \ddots \end{pmatrix}.$$

It is easy to see that B is bounded, and so B defines a bounded linear functional on $\ell_2 \hat{\otimes}_\pi \ell_2$. But

$$\sum_{i,j} |x_i x_j B(e_i, e_j)| = \sum_{n=1}^{\infty} \sum_{i,j \in I_n} |x_i x_j B(e_i, e_j)| = \sum_{n=1}^{\infty} \frac{2^{n/2}}{n^2},$$

which diverges. Therefore the series $\sum x_i x_j e_i \otimes e_j$ is only conditionally convergent.

This shows that the tensor product basis for $\ell_2 \hat{\otimes}_\pi \ell_2$ is not unconditional. It follows that the tensor product basis for the injective tensor product $\ell_2 \hat{\otimes}_\varepsilon \ell_2$ also fails to be unconditional, since $(\ell_2 \hat{\otimes}_\varepsilon \ell_2)^* = \ell_2 \hat{\otimes}_\pi \ell_2$.

4.4 Exercises

Exercise 4.1. Let X have the approximation property and let $S\colon X \to Y$ and $T\colon W \to Z$ be injective operators. Show that the operator $S \otimes_\pi T\colon X \hat{\otimes}_\pi W \to Y \hat{\otimes}_\pi Z$ is injective.

Exercise 4.2. Let $A(D)$ be the Banach space of complex functions on the closed unit disc, D, in the complex plane, that are continuous on D and analytic in the interior of D, where the norm is given by $\|f\| = \sup\{|f(z)| : z \in D\}$. Show that $A(D)$ has the approximation property.

Exercise 4.3. Show that if X^* or Y^* has the metric approximation property, then the canonical mapping of $X^*\hat{\otimes}_\pi Y^*$ into $(X\hat{\otimes}_\varepsilon Y)^*$ is an isometric embedding.

Exercise 4.4. If X^* or Y has the metric approximation property, then $\mathcal{N}(X,Y)$ is a closed subspace of $\mathcal{I}(X,Y)$. What can be said if the metric approximation property is replaced by the bounded approximation property?

Exercise 4.5. Let X and Y be Banach spaces with the bounded (respectively, metric) approximation property. Show that both $X\hat{\otimes}_\pi Y$ and $X\hat{\otimes}_\varepsilon Y$ have the bounded (respectively, metric) approximation property.

Exercise 4.6. Show that the space $\ell_3\hat{\otimes}_\pi\ell_3\hat{\otimes}_\pi\ell_3$ is not reflexive.

Exercise 4.7. Let (e_n), (f_n) be shrinking bases for X, Y respectively. Show that the tensor product basis $(e_n \otimes f_m)$ for $X\hat{\otimes}_\pi Y$ is shrinking if and only if every operator from X to Y^* is compact.

Exercise 4.8. Let (e_n), (f_n) be boundedly complete bases for X, Y respectively. Find a necessary and sufficient condition on X and Y so that the tensor product basis for $X\hat{\otimes}_\varepsilon Y$ is also boundedly complete.

Exercise 4.9. Show that every operator from c_0 into ℓ_1 is compact.

Exercise 4.10. Let X be a Banach space with an unconditional basis. Taking the standard bases for ℓ_1 and c_0, show that the tensor product bases in $\ell_1\hat{\otimes}_\pi X$ and $c_0\hat{\otimes}_\varepsilon X$ are unconditional.

5. The Radon–Nikodým Property

In this chapter we introduce the Radon–Nikodým property for Banach spaces. We begin with a study of vector measures, that is, measures with values in a Banach space. Those spaces for which the classical Radon–Nikodým Theorem extends to vector valued measures are said to have the Radon–Nikodým property. The identification of injective and projective tensor products of spaces of scalar measures in terms of spaces of vector measures sheds some light on this property. We then examine the representability of various types of operators on $C(K)$ and $L_1(\mu)$ spaces and we uncover some classes of Banach spaces, such as the reflexive spaces and the separable dual spaces, that possess the Radon–Nikodým property. We also relate the possession of this property to the coincidence of the integral and nuclear operators. Finally, we give some applications of the Radon–Nikodým property, including the Principle of Local Reflexivity.

5.1 Vector Measures and the Radon–Nikodým Property

Let Σ be a σ-algebra of subsets of a set Ω and let $\mathcal{M}(\Sigma)$ denote the Banach space of scalar measures on Σ with the variation norm, $\|\mu\| = |\mu|(\Omega)$, where $|\mu|(E)$ denotes the variation of μ on the set E.

A *vector measure* on Σ is a countably additive function μ on Σ with values in a Banach space X. Thus, if (E_n) is a sequence of mutually disjoint measurable subsets of Ω with union E, then the series $\sum_n \mu(E_n)$ is required to converge to $\mu(E)$. Since the ordering of the terms is arbitrary, it follows that this series must converge unconditionally. Furthermore, thanks to the Orlicz–Pettis Theorem (Proposition 3.12), in order to establish countable additivity for a function $\mu\colon \Sigma \to X$, it suffices to show that the scalar function $\varphi\mu$ is countable additive for every $\varphi \in X^*$. In other words, $\mu\colon \Sigma \to X$ is a vector measure if and only if $\varphi\mu$ is a scalar measure for every $\varphi \in X^*$.

We now consider the definition of the *variation* for vector measures. Proceeding exactly as in the scalar case, let

$$|\mu|_1(E) = \sup\left\{\sum_{i=1}^{n} \|\mu(A_i)\| : \{A_1, \ldots, A_n\} \text{ a partition of } E\right\},$$

for each $E \in \Sigma$, where a partition of E is a finite set of mutually disjoint measurable sets whose union is E. It is easy to see that the variation is a monotonic, non-negative, extended real-valued function. Now, the variation of a scalar measure is a finite, positive measure. However, in the vector valued case, the variation need not be finite valued.

The vector measure μ is said to have *bounded variation* if $|\mu|_1(E)$ is finite for every $E \in \Sigma$, or equivalently, if $|\mu|_1(\Omega)$ is finite. We leave it as an exercise to the reader to verify that if μ has bounded variation, then the variation $|\mu|_1$ is a finite, positive measure on Σ. Furthermore, it is not difficult to show that the set of vector measures with values in X that have bounded variation is a Banach space with respect to the *variation norm*:

$$\|\mu\|_1 = |\mu|_1(\Omega)\,.$$

We illustrate these ideas with some examples.

Example 5.1. *Some c_0-valued vector measures.*

Let Σ be the σ-algebra of all subsets of $\mathbb{N}$. Fix an element $\alpha = (\alpha_n)$ of c_0 and let

$$\mu(E) = \sum_{n \in E} \alpha_n e_n\,.$$

It is easy to see that μ is a vector measure with variation $|\mu|_1(E) = \sum_{n \in E} |\alpha_n|$. Therefore μ has bounded variation if and only if α belongs to ℓ_1.

Example 5.2. *An $L_p[0,1]$-valued vector measure.*

Let Σ be the σ-algebra of Lebesgue measurable subsets of $[0,1]$ and let $\mu\colon \Sigma \to L_p[0,1]$ be given by

$$\mu(E) = \chi_E\,.$$

If $1 \le p < \infty$, then μ is a vector measure. However, if $p = \infty$, then μ is only finitely additive. When $p = 1$, the variation of μ on a measurable set E is $m(E)$, the Lebesgue measure of E. Thus μ has bounded variation. We leave it to the reader to verify that the variation of μ is unbounded for $1 < p < \infty$.

There is an alternative approach to the definition of a variation for vector measures that has the advantage of always being bounded. Let μ be a vector measure on Σ with values in X. Then $\varphi\mu$ is a scalar measure for every $\varphi \in X^*$ and so we may define a linear mapping $T_\mu\colon X^* \to \mathcal{M}(\Sigma)$ by $T_\mu\varphi = \varphi\mu$. A straightforward application of the Closed Graph Theorem shows that T_μ is bounded. Therefore, we may define the *semivariation* of μ on a measurable set E by

$$|\mu|_\infty(E) = \sup\{|\varphi\mu|(E) : \varphi \in B_{X^*}\}\,,$$

and we see that

$$|\mu|_\infty(E) \le |\mu|_1(E) \quad \text{and} \quad |\mu|_\infty(E) \le |\mu|_\infty(\Omega) = \|T_\mu\|$$

for every $E \in \Sigma$. It follows easily from the definition that the semivariation of μ is a monotone, countably subadditive function on Σ, taking its values in a bounded subset of $\mathbb{R}$. However, the semivariation is not, in general, countably additive.

The semivariation satisfies a useful inequality:

$$\|\mu(E)\| \le |\mu|_\infty(E) \le 4\sup\{\|\mu(F)\| : F \in \Sigma, F \subset E\} \tag{5.1}$$

for every $E \in \Sigma$. This is an immediate consequence of the corresponding fact for scalar measures and the definition of the semivariation. It follows that every vector measure is bounded, since we have $\|\mu(E)\| \le |\mu|_\infty(E) \le |\mu|_\infty(\Omega) = \|T_\mu\|$ for every $E \in \Sigma$. However, we shall see shortly that a much stronger statement holds.

We denote by $\mathcal{M}(\Sigma, X)$ the vector space of vector measures on Σ with values in X. The *semivariation norm* is defined on this space by

$$\|\mu\|_\infty = |\mu|_\infty(\Omega) = \|T_\mu\|\,.$$

Proposition 5.3. *The space $\mathcal{M}(\Sigma, X)$ of vector measures is complete in the semivariation norm.*

Proof. Let (μ_n) be a Cauchy sequence in $\mathcal{M}(\Sigma, X)$. Since

$$\|\mu_n(E) - \mu_m(E)\| \le |\mu_n - \mu_m|_\infty(E) \le \|\mu_n - \mu_m\|_\infty\,,$$

the sequence $(\mu_n(E))$ is Cauchy in X for every $E \in \Sigma$. Therefore, we may define $\mu\colon \Sigma \to X$ by $\mu(E) = \lim_n \mu_n(E)$ for every $E \in \Sigma$. Then $\varphi\mu(E) = \lim_n \varphi\mu_n(E)$ for every $E \in \Sigma$, $\varphi \in X^*$, and it follows from the Nikodým Convergence Theorem (see Appendix C) that $\varphi\mu$ is a scalar measure for every $\varphi \in X^*$. Therefore μ belongs to $\mathcal{M}(\Sigma, X)$. Let $T_\mu\colon X^* \to \mathcal{M}(\Sigma)$ be the operator associated with μ, so that $T_\mu(\varphi) = \varphi\mu$ for $\varphi \in X^*$. Since $\|T_{\mu_n} - T_{\mu_m}\| = \|\mu_n - \mu_m\|_\infty$, the sequence (T_{μ_n}) is Cauchy in $\mathcal{L}(X^*, \mathcal{M}(\Sigma))$. It is easy to see that the operator T_μ is the limit of this sequence. Therefore

$$\|\mu_n - \mu\|_\infty = \|T_{\mu_n} - T_\mu\| \to 0\,,$$

and so the sequence (μ_n) converges in semivariation norm to μ. □

Vector measures can be generated by means of indefinite integrals. Recall that if the scalar function f is integrable with respect to a positive measure λ, then the formula $\mu(E) = \int_E f\,d\lambda$ defines a measure with variation given by $|\mu|_1(E) = \int_E |f|\,d\lambda$. This process can be imitated in the vector valued case, using the Pettis integral. If the function is assumed to possess stronger properties, such as measurability or Bochner integrability, we see a corresponding improvement in the behaviour of the indefinite integral:

Proposition 5.4. *Let λ be a finite, positive measure on the σ-algebra Σ.*

(a) *If $f\colon \Omega \to X$ is Pettis integrable with respect to λ, then $\mu(E) = \int_E f\, d\lambda$ defines a vector measure on Σ with semivariation given by*

$$|\mu|_\infty(E) = \sup\left\{ \int_E |\varphi f|\, d\lambda : \varphi \in B_{X^*} \right\}$$

for every $E \in \Sigma$. In particular, the semivariation norm of μ satisfies $\|\mu\|_\infty = |f|_1$, where $|f|_1$ is the Pettis norm of f.

(b) *If f is a λ-measurable Pettis integrable function, then the vector measure μ takes its values in a compact subset of X.*

(c) *If f is Bochner integrable with respect to λ, then μ has bounded variation and the variation is given by*

$$|\mu|_1(E) = \int_E \|f\|\, d\mu\,.$$

In particular, the variation norm of μ satisfies $\|\mu\|_1 = \|f\|_1$, where $\|f\|_1$ is the Bochner norm of f.

Proof. (a) By the definition of the Pettis integral, $\varphi\mu(E) = \int_E \varphi f\, d\lambda$ for $\varphi \in X^*$ and $E \in \Sigma$. Therefore $\varphi\mu$ is a measure for every $\varphi \in X^*$ and so μ is a vector measure. The formula for the semivariation of μ follows immediately from the fact that $|\varphi\mu|(E) = \int_E |\varphi f|\, d\lambda$.

(b) If f is λ-measurable and Pettis integrable, then, by Proposition 3.11, the operator $T\colon L_\infty(\lambda) \to X$, given by $Tg = \int_\Omega g f\, d\lambda$, is compact. Since $\mu(E) = T\chi_E$, it follows that the values $\mu(E)$ lie in a compact subset of X.

(c) Suppose that f is Bochner integrable with respect to λ. If $\{A_1, \ldots, A_m\}$ is a partition of a measurable set E, then

$$\sum_{i=1}^m \|\mu(A_i)\| = \sum_{i=1}^m \left\| \int_{A_i} f\, d\lambda \right\| \le \sum_{i=1}^m \int_{A_i} \|f\|\, d\lambda = \int_E \|f\|\, d\lambda\,.$$

Therefore μ has bounded variation and $|\mu|_1(E) \le \int_E \|f\|\, d\lambda$ for every $E \in \Sigma$. To prove the reverse inequality we approximate f by a simple function. Fix $\varepsilon > 0$ and let g be a λ-measurable simple function such that $\int_\Omega \|f - g\|\, d\lambda < \varepsilon$. Let ν be the vector measure given by the indefinite integral of g. It follows easily that $\big| |\mu|_1(E) - |\nu|_1(E) \big| \le \varepsilon$. Let $g = \sum_{i=1}^n \chi_{E_i} x_i$ be the canonical representation of g. Let $\{A_1, \ldots, A_m\}$ be a partition of $E \in \Sigma$. We may assume, by refining this partition, that each of the sets $E \cap E_i$ is a union of a subset of $\{A_j\}$. Then

$$\sum_{j=1}^m \|\nu(A_j)\| = \sum_{j=1}^m \left\| \int_{A_j} g\, d\lambda \right\| = \sum_{i=1}^n \|x_i\| \lambda(E \cap E_i) = \int_E \|g\|\, d\lambda$$

and hence $|\nu|_1(E) = \int_E \|g\|\, d\lambda$. Hence

$$|\mu|_1(E) \le |\nu|_1(E) + \varepsilon = \int_E \|g\|\, d\lambda + \varepsilon \le \int_E \|f\|\, d\lambda + 2\varepsilon$$

for every $\varepsilon > 0$. Therefore $|\mu|_1(E) \le \int_E \|f\|\, d\lambda$ and the proof is complete. □

A deeper understanding of vector measures requires a knowledge of the structure of the weakly compact subsets of the Banach space $\mathcal{M}(\Sigma)$ of scalar measures. For the benefit of readers who may be unfamiliar with this material, a complete account is given in Appendix C. In particular, we shall make use of the following result.

Proposition 5.5. *Let K be a bounded subset of $\mathcal{M}(\Sigma)$. The following statements are equivalent:*

(i) *K is relatively weakly compact.*
(ii) *The elements of K are uniformly countably additive.*
(iii) *There exists a finite, positive measure λ on Σ such that the elements of K are uniformly absolutely continuous with respect to λ.*

We recall that to say that the elements of K are uniformly countably additive means that if (E_n) is a sequence of mutually disjoint measurable sets, then the convergence of the series $\sum_{n=1}^{\infty} \mu(E_n)$ is uniform in $\mu \in K$. Equivalently, if (F_n) is a decreasing sequence of measurable sets with intersection F, then $\mu(F_n)$ tends to $\mu(F)$ uniformly in $\mu \in K$. The statement that the elements of K are uniformly absolutely continuous with respect to λ means that, for every $\varepsilon > 0$ there exists $\delta > 0$ such that if E is a measurable set with $\lambda(E) < \delta$, then $|\mu(E)| < \varepsilon$ for every $\mu \in K$.

Now let $\mu\colon \Sigma \to X$ be a vector measure and let K be the bounded subset of $\mathcal{M}(\Sigma)$ consisting of the scalar measures $\varphi\mu$, where $\varphi \in B_{X^*}$. If (E_n) is a sequence of mutually disjoint measurable sets, then

$$\left| \sum_{k=n}^{\infty} \varphi\mu(E_k) \right| \le \left\| \sum_{k=n}^{\infty} \mu(E_k) \right\|$$

for every $\varphi \in B_{X^*}$. It follows that the measures $\varphi\mu$, as φ ranges over B_{X^*}, are uniformly countably additive and hence K is relatively weakly compact in $\mathcal{M}(\Sigma)$. In terms of the operator associated with μ, we have:

Proposition 5.6. *Let $\mu\colon \Sigma \to X$ be a vector measure. Then the operator $T_\mu\colon X^* \to \mathcal{M}(\Sigma)$, given by $T_\mu(\varphi) = \varphi\mu$, is weakly compact.*

We have seen that the range of a vector measure is bounded. In fact, much more is true:

Corollary 5.7 (Bartle–Dunford–Schwartz Theorem). *Let μ be a vector measure. Then the values of μ lie in a weakly compact set.*

Proof. Consider the weakly compact adjoint operator $T_\mu^*\colon \mathcal{M}(\Sigma)^* \to X^{**}$. For each $E \in \Sigma$, the characteristic function χ_E acts as a bounded linear functional on the space $\mathcal{M}(\Sigma)$ by $\langle \lambda, \chi_E \rangle = \lambda(E)$ and has unit norm. It is easy to see that $\mu(E) = T_\mu^*(\chi_E)$. Therefore the subset $K = \{\mu(E) : E \in \Sigma\}$ of X is contained in the relatively weakly compact subset $T_\mu^*(B)$ of X^{**}, where B is the closed unit ball of $\mathcal{M}(\Sigma)^*$, and it follows that K is relatively weakly compact in X. □

The operator $T_\mu\colon X^* \to \mathcal{M}(\Sigma)$ associated with a vector measure can be thought of as an "adjoint" of μ, with measure-theoretic properties of μ being reflected by operator-theoretic properties of T_μ. Our next result is another illustration of this principle.

Proposition 5.8. *Let $\mu\colon \Sigma \to X$ be a vector measure with the associated operator $T_\mu\colon X^* \to \mathcal{M}(\Sigma)$ given by $T_\mu(\varphi) = \varphi\mu$. Then the following are equivalent:*

(i) *μ has bounded variation.*
(ii) *T_μ is a Pietsch integral operator.*
(iii) *T_μ is an integral operator.*

Furthermore, if μ has bounded variation, then

$$\|\mu\|_1 = \|T_\mu\|_I = \|T_\mu\|_{PI}\,.$$

Proof. (i) implies (ii): suppose that μ has bounded variation. Let λ denote the variation of μ. Then λ is a finite, positive measure on Σ. We shall show that there is an operator $V\colon X^* \to L_\infty(\lambda)$ of unit norm, such that T_μ factors as UIV, where $U\colon L_1(\lambda) \to \mathcal{M}(\Sigma)$ is the canonical embedding given by $Uf(E) = \int_E f\,d\lambda$ and $I\colon L_\infty(\lambda) \to L_1(\lambda)$ is the canonical injection:

$$\begin{array}{ccc} X^* & \xrightarrow{\ T_\mu\ } & \mathcal{M}(\Sigma) \\ {\scriptstyle V}\big\downarrow & & \big\uparrow{\scriptstyle U} \\ L_\infty(\lambda) & \xrightarrow{\ \ I\ \ } & L_1(\lambda) \end{array}$$

It then follows that T_μ is Pietsch integral, with Pietsch integral norm satisfying

$$\|T_\mu\|_{PI} \le \|\lambda\| = \|\mu\|_1\,.$$

To define the operator V, let $\varphi \in X^*$. The measure $\varphi\mu$ is absolutely continuous with respect to λ since, if $\lambda(E) = 0$, then $\mu(E) = 0$ and hence $\varphi\mu(E) = 0$. Therefore, by the Radon–Nikodým Theorem, there exists a unique element f_φ of $L_1(\lambda)$ such that $\varphi\mu(E) = \int_E f_\varphi\,d\mu$ for every $E \in \Sigma$. The mapping $\varphi \in X^* \mapsto f_\varphi \in L_1(\lambda)$ defines an operator and we have

$$T_\mu(\varphi)(E) = \varphi\mu(E) = \int_E f_\varphi \, d\mu = Uf_\varphi(E)$$

for every $E \in \Sigma$. Thus $T_\mu \varphi = U f_\varphi$ for every $\varphi \in X^*$.

We claim that $f_\varphi \in L_\infty(\lambda)$ for every $\varphi \in X^*$. To see this, let g be a measurable, simple function with canonical representation $\sum_{i=1}^n c_i \chi_{E_i}$ and let $\varphi \in X^*$. Then

$$\left| \int_\Omega g f_\varphi \, d\lambda \right| = \left| \sum_{i=1}^n c_i \varphi\mu(E_i) \right| \le \|\varphi\| \sum_{i=1}^n |c_i| \|\mu(E_i)\|$$
$$\le \|\varphi\| \sum_{i=1}^n |c_i| \lambda(E_i) = \|\varphi\| \|g\|_1 .$$

It follows that f_φ belongs to $L_\infty(\lambda)$ for every $\varphi \in X^*$ and that the operator $V : \varphi \in X^* \mapsto f_\varphi \in L_\infty(\lambda)$ has norm at most one. Finally, we have $T_\mu = UIV$ as asserted.

(ii) implies (iii) is trivial, so we proceed to the proof of (iii) implies (i). Suppose that the operator T_μ is integral. Then T_μ defines a bounded linear functional on the injective tensor product $X^* \hat{\otimes}_\varepsilon \mathcal{M}(\Sigma)^*$. Let E be a measurable subset of Ω and let $\{A_1, \dots, A_n\}$ be a partition of E. For each i, let $\varphi_i \in B_{X^*}$ be such that $\|\mu(A_i)\| = \langle \mu(A_i), \varphi_i \rangle$. Then

$$\sum_{i=1}^n \|\mu(A_i)\| = \sum_{i=1}^n \langle \mu(A_i), \varphi_i \rangle = \sum_{i=1}^n \langle T_\mu \varphi_i, \chi_{A_i} \rangle$$
$$= \left\langle \sum_{i=1}^n \varphi_i \otimes \chi_{A_i}, T_\mu \right\rangle \le \varepsilon \left(\sum_{i=1}^n \varphi_i \otimes \chi_{A_i} \right) \|T_\mu\|_I .$$

We compute the injective tensor norm:

$$\varepsilon \left(\sum_{i=1}^n \varphi_i \otimes \chi_{A_i} \right) = \sup_{x \in B_X} \left\| \sum_{i=1}^n \varphi_i(x) \chi_{A_i} \right\|_{\mathcal{M}(\Sigma)^*}$$
$$= \sup_{x \in B_X} \sup_{\|\nu\|=1} \left| \sum_{i=1}^n \varphi_i(x) \nu(A_i) \right| \le \sup_{\|\nu\|=1} \sum_{i=1}^n |\nu(A_i)| \le 1 .$$

Therefore $\sum_{i=1}^n \|\mu(A_i)\| \le \|T_\mu\|_I$ for every partition of E and so μ has bounded variation. Furthermore, we have $\|\mu\|_1 \le \|T_\mu\|_I$. Therefore

$$\|\mu\|_1 \le \|T\mu\|_I \le \|T_\mu\|_{PI} \le \|\mu\|_1 ,$$

and so these norms coincide. □

We now consider absolute continuity for vector measures. The definition is the same as in the scalar case: the vector measure μ is said to be *absolutely continuous* with respect to the finite, positive measure λ if, for every $\varepsilon > 0$ there exists $\delta > 0$ such that if E is a measurable set with $\lambda(E) < \delta$, then $\|\mu(E)\| < \varepsilon$. As in the scalar case, we have:

Proposition 5.9. *Let $\mu\colon \Sigma \to X$ be a vector measure and let λ be a finite, positive measure on Σ. Then μ is absolutely continuous with respect to λ if and only if $\lambda(E) = 0$ implies $\mu(E) = 0$.*

Proof. It is clear that this condition is implied by absolute continuity. Conversely, suppose that μ is not absolutely continuous with respect to λ. Then there exists $\varepsilon > 0$ and a sequence (E_n) of measurable sets such that $\lambda(E_n) < 2^{-n}$ and $|\mu|_\infty(E_n) \geq \|\mu(E_n)\| \geq \varepsilon$ for every n. Let $F_n = \bigcup_{k>n} E_k$ and let F be the intersection of the decreasing sequence (F_n). Then $\lambda(F) = \lim_n \lambda(F_n) = 0$. Since the set $\{|\varphi\mu| : \varphi \in B_{X^*}\}$ is weakly compact, $|\varphi\mu|(F_n)$ converges to $|\varphi\mu|(F)$ uniformly for $\varphi \in B_{X^*}$. Therefore $|\mu|_\infty(F) \geq \varepsilon$. Now, by (5.1), F contains a measurable subset G with $\|\mu(G)\| \geq |\mu|_\infty(F)/4$. Thus $\lambda(G) = 0$, but $\mu(G) \neq 0$. □

We now construct an integral for scalar functions with respect to a vector measure. It will be sufficient for the applications that we have in mind to define the integral for bounded measurable functions. While it is possible to define the integral first for simple functions and then pass to limits of sequences of simple functions, we take a different approach that enables us to make use of some of the machinery that we have developed. Let $B(\Omega, \Sigma)$ denote the Banach space of bounded scalar functions on Ω that are measurable with respect to the σ-algebra Σ, with the norm $\|f\|_\infty = \sup\{|f(\omega)| : \omega \in \Omega\}$. We note that the Σ-measurable simple functions are dense in $B(\Omega, \Sigma)$. The space $B(\Omega, \Sigma)$ can be considered as a subspace of the dual space $\mathcal{M}(\Sigma)^*$, where $\langle \lambda, f \rangle = \int_\Omega f\, d\lambda$ for $\lambda \in \mathcal{M}(\Sigma)$, $f \in B(\Omega, \Sigma)$.

Now let $\mu\colon \Sigma \to X$ be a vector measure and let $T_\mu\colon X^* \to \mathcal{M}(\Sigma)$ be the associated operator. Restricting the adjoint of T_μ to the space $B(\Omega, \Sigma)$ yields an operator $S\colon B(\Omega, \Sigma) \to X^{**}$. We have $S\chi_E = \mu(E)$ for every $E \in \Sigma$ and so, if $g = \sum_{i=1}^n c_i \chi_{E_i}$ is the canonical representation of a Σ-measurable simple function, then $Sg = \sum_{i=1}^n c_i \mu(E_i) \in X$. Since these functions are dense in $B(\Omega, \Sigma)$, it follows that S takes its values in X. Thus, we may define the integral of a bounded, Σ-measurable function f over a measurable set E by $\int_E f\, d\mu = S(\chi_E f) \in X$. The following properties of this integral are immediate.

Proposition 5.10. *Let $\mu\colon \Sigma \to X$ be a vector measure.*

(a) *If f is a bounded Σ-measurable function on Ω and $E \in \Sigma$ then*

$$\left\| \int_E f\, d\mu \right\| \leq \|f\|_\infty |\mu|_\infty(E)\,.$$

In particular,

$$\left\| \int_\Omega f\, d\mu \right\| \leq \|f\|_\infty \|\mu\|_\infty\,.$$

$$\mu(E) = \int_E f \, d\lambda$$

has the Radon–Nikodým property.

Proof. By the Pettis Measurability Theorem, we may assume that X is separable. Consider the function $h\colon X \to X$ defined by

$$h(x) = \begin{cases} \|x\|^{-1}x & \text{if } x \neq 0, \\ 0 & \text{if } x = 0. \end{cases}$$

This function is Borel measurable and so, again using the Pettis Measurability Theorem, the function $g = h \circ f$ is $|\mu|_1$-measurable. Furthermore, g is bounded, and hence is Bochner integrable with respect to $|\mu|_1$. Now, the variation of μ is given by

$$|\mu|_1(E) = \int_E \|f\| \, d\lambda$$

and so we have

$$\mu(E) = \int_E f \, d\lambda = \int_E g\|f\| \, d\lambda = \int_E g \, d|\mu|_1 \, .$$

It follows from the remarks preceding the lemma that μ has the Radon–Nikodým property. □

We shall denote by $\mathcal{M}_1(\Sigma, X)$ the space of measures with the Radon–Nikodým property, with the norm given by the variation.

Lemma 5.21. *The space $\mathcal{M}_1(\Sigma, X)$ is complete in the variation norm.*

Proof. Let (μ_n) be a Cauchy sequence in $\mathcal{M}_1(\Sigma, X)$. Let λ be a finite, positive measure on Σ such that each of the scalar measures $|\mu_n|_1$ is absolutely continuous with respect to λ. Then there is a sequence (f_n) in $L_1(\lambda, X)$ such that

$$\mu_n(E) = \int_E f_n \, d\lambda$$

for every $E \in \Sigma$ and every n. Then (f_n) is Cauchy in $L_1(\lambda, X)$, since

$$\int_\Omega \|f_n - f_m\| \, d\lambda = \|\mu_n - \mu_m\|_1$$

for all m, n. Let f be the limit of this sequence. Then the sequence (μ_n) converges in variation norm to the measure $\mu(E) = \int_E f \, d\lambda$ and this measure, being given by an indefinite Bochner integral, has the Radon–Nikodým property. □

We can now characterize the projective tensor product $\mathcal{M}(\Sigma)\hat{\otimes}_\pi X$.

Theorem 5.22. *Let Σ be a σ-algebra of subsets of a set Ω and let X be a Banach space. The projective tensor product $\mathcal{M}(\Sigma)\hat{\otimes}_\pi X$ is isometrically isomorphic to the Banach space $\mathcal{M}_1(\Sigma, X)$ of vector measures with the Radon–Nikodým property, with the variation norm.*

Proof. We have seen that the projective norm coincides with the variation norm. Therefore it only remains to show that $\mathcal{M}(\Sigma) \otimes X$ is dense in $\mathcal{M}_1(\Sigma, X)$. Let $\mu \in \mathcal{M}_1(\Sigma, X)$. Then there exists $f \in L_1(|\mu|_1, X)$ such that $\mu(E) = \int_E f\, d|\mu|_1$ for every $E \in \Sigma$. Since $L_1(|\mu|_1, X) = L_1(|\mu|_1)\hat{\otimes}_\pi X$, there exist bounded sequences (g_n) and (x_n) in $L_1(|\mu|_1)$ and X respectively, such that $\sum_{n=1}^\infty \|g_n\|\|x_n\| < \infty$ and $f = \sum_{n=1}^\infty g_n \otimes x_n$, the series converging in $L_1(|\mu|_1)\hat{\otimes}_\pi X$. Let $\mu_n \in \mathcal{M}(\Sigma)$ be defined by $\mu_n(E) = \int_E g_n\, d|\mu|_1$. Then $\mu = \sum_{n=1}^\infty \mu_n \otimes x_n \in \mathcal{M}(\Sigma)\hat{\otimes}_\pi X$. □

Corollary 5.23. *If X has the Radon–Nikodým property then $\mathcal{M}(\Sigma)\hat{\otimes}_\pi X$ is the space of vector measures of bounded variation, with the variation norm.*

5.3 Operators on $C(K)$ Spaces

Let K be a compact topological space and let $\mathcal{B}_K$ be the σ-algebra of Borel sets. Every bounded linear functional on the space $C(K)$ is given by integration with respect to a uniquely determined regular measure on the σ-algebra $\mathcal{B}_K$. It is natural to ask if there is an integral representation for operators. In other words, if $T\colon C(K) \to X$ is an operator, does there exist a vector measure μ with values in X such that

$$Tf = \int_K f\, d\mu$$

for every $f \in C(K)$?

Let us see how we might attempt to construct such a vector measure. The dual space $C(K)^*$ is the Banach space of regular Borel measures on K with the variation norm. The Banach space of bounded Borel measurable functions on K, with the supremum norm, is a subspace of $C(K)^{**}$; if g is such a function and $\mu \in C(K)^*$, then $\langle \mu, g\rangle = \int_K g\, d\mu$. In particular, we may consider the characteristic functions of Borel subsets of K as elements of $C(K)^{**}$. Now let $T\colon C(K) \to X$ be an operator. We may define a function μ_T on the σ-algebra of Borel subsets of K with values in the bidual X^{**} by

$$\mu_T(E) = T^{**}(\chi_E)\,. \tag{5.2}$$

Thus, for each $\varphi \in X^*$, we have

$$\langle \varphi, \mu_T(E)\rangle = T^*\varphi(E)\,, \tag{5.3}$$

where $T^*\varphi \in C(K)^*$ is interpreted as a regular Borel measure on K. It is unfortunate that μ_T takes its values in X^{**} rather than X. However, this function has almost all the properties we require. To begin with, we see that μ_T is *weak*-countably additive*: if (E_n) is a sequence of mutually disjoint Borel subsets of K, then it follows from (5.3) that the series $\sum_n \mu_T(E_n)$ is weak*-convergent to $\mu_T(E)$. Furthermore, μ_T is *weak*-regular* in the sense that the measure $\varphi\mu_T$ is regular for every $\varphi \in X^*$. And finally, we have an integral representation of sorts:

$$\langle Tf, \varphi\rangle = \int_K f \, d\varphi\mu_T \tag{5.4}$$

for every $f \in C(K)$ and every $\varphi \in X^*$. We shall refer to the function μ_T as the *representing measure* for the operator T. It is clear from (5.2) that μ_T will take its values in X if T^{**} maps $C(K)^{**}$ into X, in other words, if T is a weakly compact operator. Before developing this idea, we clarify the meaning of regularity for vector measures. Just as in the scalar case, a vector measure μ on the σ-algebra of Borel subsets of K is said to be *regular* if, for every Borel set E and every $\varepsilon > 0$, there exist an open set O containing E and a closed set C contained in E, such that $\|\mu(O \setminus C)\| < \varepsilon$. Our next result shows that regularity is equivalent to "weak regularity".

Lemma 5.24. *Let μ be a vector measure on the σ-algebra of Borel subsets of the compact space K with values in the Banach space X. Then μ is regular if and only if the scalar measure $\varphi\mu$ is regular for every $\varphi \in X^*$.*

Proof. It is easy to see that the regularity of μ implies that of $\varphi\mu$ for every $\varphi \in X^*$. Conversely, suppose that $\varphi\mu$ is regular for every $\varphi \in X^*$. Since the set $\{\varphi\mu : \varphi \in B_{X^*}\}$ is weakly compact in the space of regular Borel measures, it follows that there is a regular positive Borel measure λ such that the measures in this set are uniformly absolutely continuous with respect to λ (see Appendix C). Thus, given $\varepsilon > 0$, there exists $\delta > 0$ such that if E is a Borel set with $\lambda(E) < \delta$, then $|\varphi\mu(E)| < \varepsilon$ for every $\varphi \in B_{X^*}$, and hence $\|\mu(E)\| < \varepsilon$. The regularity of μ now follows immediately from the regularity of λ. □

We can now present the fundamental result about the correspondence between weakly compact operators on $C(K)$ and vector measures on the Borel σ-algebra of K.

Theorem 5.25. *Let $T\colon C(K) \to X$ be a weakly compact operator. Then there exists a unique regular vector measure μ on the Borel subsets of K with values in X such that*

$$Tf = \int_K f \, d\mu$$

for every $f \in C(K)$. Conversely, if μ is a regular vector measure on the Borel subsets of K with values in X, then $Tf = \int_K f \, d\mu$ defines a weakly compact operator from $C(K)$ into X. If T and μ are related in this way, then

$$\|T\| = \|\mu\|_\infty ,$$

the semivariation norm of μ.

Proof. We have seen that $\mu(E) = T^{**}(\chi_E)$ defines a weak*-countably additive, weak*-regular function from the Borel σ-algebra $\mathcal{B}_K$ into X^{**}, and that

$$\langle Tf, \varphi\rangle = \int_K f \, d\varphi\mu$$

for every $f \in C(K)$ and $\varphi \in X^*$. If T is weakly compact, then T^{**} maps $(C(K))^{**}$ into X and hence $\mu(E) \in X$ for every Borel set E. It follows that μ is weakly countably additive and so μ is a vector measure. Furthermore, the scalar measure $\varphi\mu$ is regular for every $\varphi \in X^*$ and so μ is regular by Lemma 5.24. Let $f \in C(K)$. Since

$$\langle Tf, \varphi\rangle = \int_K f \, d\varphi\mu = \left\langle \int_K f \, d\mu, \varphi \right\rangle$$

for every $\varphi \in X^*$, it follows that $Tf = \int_K f \, d\mu$.

Conversely, let μ be a regular vector measure on the Borel subsets of K with values in X and let $T\colon C(K) \to X$ be the operator defined by $Tf = \int_K f \, d\mu$. Then $T^*\varphi$ is the scalar measure $\varphi\mu$ for $\varphi \in X^*$. Since $\{\varphi\mu : \varphi \in B_{X^*}\}$ is a relatively weakly compact set of measures, it follows that T^*, and hence T, is a weakly compact operator.

Now

$$\|Tf\| = \left\| \int_K f \, d\mu \right\| \leq \|f\| \|\mu\|_\infty$$

for every $f \in C(K)$, and so $\|T\| \leq \|\mu\|_\infty$. On the other hand,

$$\|\varphi\mu\| = \|T^*\varphi\| \leq \|T^*\| = \|T\|$$

for every $\varphi \in B_{X^*}$, and hence $\|\mu\|_\infty \leq \|T\|$. Therefore $\|T\| = \|\mu\|_\infty$. □

A Banach space X is said to have the *Dunford–Pettis property* if, for every Banach space Y, every weakly compact operator from X into Y is completely continuous. Now the weakly convergent sequences in $C(K)$ are precisely the uniformly bounded pointwise convergent sequences. Therefore, the above proposition in tandem with the Bounded Convergence Theorem for vector measures (Proposition 5.12) yields:

Corollary 5.26. *The space $C(K)$ has the Dunford–Pettis property.*

Properties of the operator $T\colon C(K) \to X$ are reflected by properties of the representing measure. We shall pursue this idea at some length. We begin with compactness.

Proposition 5.27. *The operator $T\colon C(K) \to X$ is compact if and only if the representing measure of T is a compact vector measure.*

Proof. Suppose that T is compact. Then so is T^{**} and, since the characteristic functions of Borel sets have unit norm in $C(K)^{**}$, the set $\{\mu(E) : E \in \mathcal{B}_K\} = \{T^{**}\chi_E\}$ is relatively compact in X. Therefore μ is compact.

Conversely, suppose that μ is a compact vector measure. By Theorem 5.18, there exists a sequence (μ_n) in $\mathcal{M}(\mathcal{B}_K) \otimes X$ that converges to μ in the semivariation norm. For each n, let $T_n : C(K) \to X$ be the finite rank operator defined by $T_n f = \int_K f \, d\mu_n$. We have

$$\|T_n f - Tf\| = \left\| \int_K f \, d(\mu_n - \mu) \right\| \leq \|f\| \, \|\mu_n - \mu\|_\infty ,$$

and it follows that the sequence (T_n) converges to T in the operator norm. Therefore T is compact. □

We now consider the class of integral operators. We have seen in Proposition 5.8 that the operator $T : X^* \to \mathcal{M}(\Sigma)$ associated with a vector measure μ is integral, or Pietsch integral, if and only if μ has bounded variation. Our next result is a variation on this theme.

Proposition 5.28. *Let $T : C(K) \to X$ be an operator with representing measure μ. The following are equivalent:*

(i) *T is a Pietsch integral operator.*
(ii) *T is an integral operator.*
(iii) *μ is a vector measure of bounded variation.*

Furthermore, if T is Pietsch integral, then

$$\|T\|_{PI} = \|T\|_I = \|\mu\|_1 ,$$

the variation norm of μ.

Proof. (i) ⇒ (ii) being trivial, we consider the implication (ii) ⇒ (iii). Suppose that T is an integral operator. Then $T^* : X^* \to C(K)^*$ is integral, with the same integral norm. Let I be the canonical inclusion of $C(K)^*$ into the space $\mathcal{M}(\mathcal{B}_K)$ of Borel measures on K. Then $IT^* : X^* \to \mathcal{M}(\mathcal{B}_K)$ is the operator associated with the vector measure μ and so, by Proposition 5.8, μ has bounded variation and

$$\|\mu\|_1 = \|IT^*\|_I \leq \|T^*\|_I = \|T\|_I .$$

To prove (iii) ⇒ (i), suppose that μ has bounded variation. Then the variation $|\mu|_1$ is a finite, positive Borel measure on K. We shall show that T is Pietsch integral by factoring it through the canonical injection $J : C(K) \to L_1(|\mu|_1)$. We define an operator $R : L_1(|\mu|_1) \to X$ as follows. Let $f = \sum_{i=1}^n \alpha_i \chi_{A_i}$ be a Borel measurable simple function, where the sets $A_1, \ldots, A_n$ are mutually disjoint. Let

$$Rf = \int_K f\,d\mu = \sum_{i=1}^{n} \alpha_i \mu(A_i)\,.$$

Then

$$\|Rf\| \le \sum_{i=1}^{n} |\alpha_i| \|\mu(A_i)\| \le \sum_{i=1}^{n} |\alpha_i| |\mu|_1(A_i) = \int_K |f|\,d|\mu|_1\,,$$

and hence R extends uniquely to an operator of unit norm. This operator satisfies

$$Rf = \int_K f\,d\mu$$

for every $f \in C(K)$. Therefore $T = RJ$ and it follows that T is Pietsch integral, with Pietsch integral norm $\|T\|_{PI} \le \|\mu\|_1$.

This proves the equivalence of (i), (ii) and (iii). Combining the inequalities obtained in the course of the proof gives $\|T\|_{PI} = \|T\|_I = \|\mu\|_1$. □

We note in passing that this result yields a representation of the dual of the space $C(K, X)$. Since this space is the injective tensor product $C(K)\hat{\otimes}_\varepsilon X$, its dual is the space of integral operators from $C(K)$ into X^*. Therefore $C(K, X)^*$ can be identified with the space of regular vector measures of bounded variation on the Borel subsets of K with values in X^*, this space being endowed with the variation norm. A complete description of this duality would require a treatment of integrals of the form $\int_K f\,d\mu$, where f takes its values in X and the vector measure μ is X^*-valued. This integral is known as the *Bartle integral.*

We conclude this treatment of operators on $C(K)$ with a characterization of the class of nuclear operators. We shall see that the operator $T\colon C(K) \to X$ is nuclear if and only if the representing measure of T has the Radon–Nikodým property. This is essentially a reformulation of Theorem 5.22, which can be interpreted as saying vector measure $\mu \in \mathcal{M}(\Sigma, X)$ has the Radon–Nikodým property if and only if the associated operator $T_\mu\colon X^* \to \mathcal{M}(\Sigma)$ is nuclear. First, we clarify the relationship between the space $\mathcal{M}(\mathcal{B}_K)$ of measures on the Borel σ-algebra of K and the subspace $C(K)^*$ of regular measures.

Lemma 5.29. *Let $\mathcal{B}_K$ be the σ-algebra of Borel subsets of the compact space K. Then $C(K)^*$ is complemented in $\mathcal{M}(\mathcal{B}_K)$ by a projection of norm one.*

Proof. Each measure $\mu \in \mathcal{M}(\mathcal{B}_K)$ defines a bounded linear functional φ on $C(K)$ by $\langle f, \varphi\rangle = \int_K f\,d\mu$. Let μ' be the unique regular measure associated with φ. Then $P\mu = \mu'$ defines a projection of $\mathcal{M}(\mathcal{B}_K)$ onto $C(K)^*$, and it is easy to see that $\|\mu'\| \le \|\mu\|$ for every μ. □

It follows that the nuclearity of an operator with values in $C(K)^*$, and the value of the nuclear norm, are unaffected if we consider the operator as taking its values in the larger space $\mathcal{M}(\mathcal{B}_K)$.

Let ν be a vector measure on a σ-algebra Σ of subsets of a set Ω with values in the Banach space X and suppose that ν has bounded variation. Then the variation $|\nu|_1$ of ν is a finite, positive measure on Σ. We have seen in the proof of Proposition 5.28 that the integral $\int_\Omega f\,d\nu$ can be defined for every $f \in L_1(|\nu|_1)$ and that it satisfies $\|\int_\Omega f\,d\nu\| \le \|f\|_1$. Thus an operator $T\colon L_1(|\nu|_1) \to X$ can be defined by

$$Tf = \int_\Omega f\,d\nu\,.$$

We now describe a property of the operator T that is equivalent to the possession by ν of the Radon–Nikodým property.

Let μ be a finite, positive measure on Σ and let X be a Banach space. An operator $T\colon L_1(\mu) \to X$ is *representable* if there exists a function $g\colon \Omega \to X$ that is μ-measurable and μ-essentially bounded, such that Tf is given by the Bochner integral $\int_\Omega fg\,d\mu$ for every $f \in L_1(\mu)$. Of course, when X is the scalar field, every operator is representable.

Proposition 5.35. *The Banach space X has the Radon–Nikodým property if and only if, for every finite, positive measure μ, every operator from $L_1(\mu)$ into X is representable.*

Proof. Suppose that, for every finite, positive measure μ, every operator from $L_1(\mu)$ into X is representable. Let $\nu\colon \Sigma \to X$ be a vector measure of bounded variation. Then the operator $T\colon L_1(|\nu|_1) \to X$ given by $Tf = \int_\Omega f\,d\nu$ is representable. Therefore there exists a $|\nu|_1$-measurable, $|\nu|_1$-essentially bounded function $g\colon \Omega \to X$ such that $Tf = \int_\Omega fg\,d|\nu|_1$ for every $f \in L_1(|\nu|_1)$. Then $\nu(E) = T\chi_E$ is given by the Bochner integral $\int_E g\,d|\nu|_1$ for every $E \in \Sigma$, and so ν has the Radon–Nikodým property.

Conversely, suppose that X has the Radon–Nikodým property and let T be an operator from $L_1(\mu)$ into X. Then $\nu(E) = T(\chi_E)$ defines a vector measure of bounded variation that is absolutely continuous with respect to μ. Indeed, we have $|\nu|_1(E) \le \|T\|\mu(E)$ for every $E \in \Sigma$. Thus there exists $g \in L_1(\mu, X)$ such that

$$\nu(E) = \int_E g\,d\mu$$

for every $E \in \Sigma$. Let us suppose for the moment that the function g is μ-essentially bounded. Then, by Bochner's Theorem, the function fg is Bochner integrable with respect to μ for every $f \in L_1(\mu)$, and we may define an operator from $L_1(\mu)$ into X by $f \mapsto \int_\Omega fg\,d\mu$. However, this operator coincides with T on the characteristic functions and, since these functions span a dense subspace of $L_1(\mu)$, it follows that

$$Tf = \int_\Omega fg\,d\mu$$

for every $f \in L_1(\mu)$. Therefore T is representable.

It only remains to show that g is μ-essentially bounded. By the Pettis Measurability Theorem, we may assume without loss of generality that X is separable. Hence there exists a countable norming subset $\{\varphi_n\}$ of B_{X^*}. Fixing n, we have

$$\int_E |\varphi_n g|\, d\mu = |\varphi_n \nu|(E) \le |\nu|_\infty(E) \le \|T\|\mu(E)$$

for every $E \in \Sigma$. It follows that there exists a μ-null set E_n such that $|\varphi_n g(\omega)| \le \|T\|$ for every $\omega \in E_n^c$. Then $|\varphi_n g(\omega)| \le \|T\|$ for every n and every ω outside the μ-null set $\bigcup_k E_k$, and so $\|g(\omega)\| \le \|T\|$ almost everywhere. □

Our next result indicates that the space ℓ_1 plays a central role in the development of the Radon–Nikodým property.

Proposition 5.36 (Lewis–Stegall Theorem). *An operator from $L_1(\mu)$ into a Banach space X is representable if and only if it factors through ℓ_1. Thus X has the Radon–Nikodým property if and only if, for every finite, positive measure μ, every operator from $L_1(\mu)$ into X factors through ℓ_1.*

Proof. Suppose that the operator $T\colon L_1(\mu) \to X$ factors through ℓ_1, so that there exists operators $S\colon L_1(\mu) \to \ell_1$ and $R\colon \ell_1 \to X$ such that $T = RS$. Since ℓ_1 has the Radon–Nikodým property, S is representable and it follows easily that T is also representable.

Conversely, suppose that the operator $T\colon L_1(\mu) \to X$ is representable, so that there exists a μ-measurable, μ-essentially bounded function g such that

$$Tf = \int_\Omega fg\, d\mu$$

for every $f \in L_1(\mu)$. We may assume without loss of generality that X is separable and that g is bounded. By the proof of the Pettis Measurability Theorem, there exists a sequence (f_n) of μ-measurable functions, with values in a countable subset of X, that converges uniformly to g. We may choose a subsequence (f_{n_k}) of this sequence such that $\|g - f_{n_k}\|_\infty < 2^{-k-1}$ for every k. Let $g_1 = f_{n_1}$ and, for $k > 1$, let $g_k = f_{n_k} - f_{n_{k-1}}$. Then g is the sum in $L_1(\mu)$ of the absolutely convergent series $\sum_{k=1}^\infty g_k$ and so

$$Tf = \sum_{k=1}^\infty \int_\Omega f g_k\, d\mu\,,$$

with this series converging absolutely in X for every $f \in L_1(\mu)$. The values of the functions g_k lie in a countable subset of X and so, for each k there exist a sequence $(x_{kj})_j$ of non-zero points in X and a sequence $(E_{kj})_j$ of mutually disjoint measurable sets, such that

$$g_k = \sum_{j=1}^{\infty} x_{kj}\chi_{E_{kj}} .$$

For each $f \in L_1(\mu)$, let

$$f_{kj} = \|x_{kj}\| \int_{E_{kj}} f \, d\mu .$$

We claim that the double sequence (f_{kj}) is summable. Indeed,

$$\begin{aligned}\sum_{k,j} |f_{kj}| \le \sum_{k=1}^{\infty}\sum_{j=1}^{\infty} \|x_{kj}\| \int_{E_{kj}} |f| \, d\mu &= \sum_{k=1}^{\infty} \int_{\Omega} |f| \|g_k\| \, d\mu \\ &\le \|f\|_1 \sum_{k=1}^{\infty} \|g_k\|_{\infty} .\end{aligned}$$

Therefore, an operator $S\colon L_1(\mu) \to \ell_1(\mathbb{N}^2)$ can be defined by $Sf = (f_{kj})$. If we define the operator $R\colon \ell_1(\mathbb{N}^2) \to X$ by

$$Ra = \sum_{k,j} a_{kj} \|x_{kj}\|^{-1} x_{kj} ,$$

then $T = RS$. Since the space $\ell_1(\mathbb{N}^2)$ is isometrically isomorphic to ℓ_1, the proof is complete. □

We note that this result can also be formulated in terms of vector measures: a vector measure has the Radon–Nikodým property if and only if it can be factored through an ℓ_1-valued measure of bounded variation.

The Lewis–Stegall Theorem can be used to infer structural properties of representable operators from special properties possessed by ℓ_1. For example, ℓ_1 has the Schur property, so that every weakly convergent sequence is convergent in norm. Hence, we have:

Corollary 5.37. *Every representable operator on $L_1(\mu)$ is completely continuous.*

We now begin to identify some classes of representable operators. We begin with compact operators.

Proposition 5.38. *Let μ be a finite, positive measure and let X be a Banach space. Then every compact operator from $L_1(\mu)$ into X is representable.*

Proof. Let $T\colon L_1(\mu) \to X$ be compact. Since $L_\infty(\mu)$ has the approximation property, T is the limit in operator norm of a sequence (T_n) of finite rank operators. Since finite rank operators are clearly representable, there exists a sequence (g_n) of μ-measurable, μ-essentially bounded functions such that $T_n f = \int_\Omega f g_n \, d\mu$ for every $f \in L_1(\mu)$. Then $\|g_n - g_m\|_\infty = \|T_n - T_m\|$ for every n, m and it follows that (g_n) converges in the essential supremum norm to a μ-measurable, μ-essentially bounded function g. An application of the Dominated Convergence Theorem for Bochner integrals shows that

$$Tf = \int_\Omega fg\,d\mu$$

for every $f \in L_1(\mu)$. Therefore f is representable. □

We can extend this result to weakly compact operators with separable range by means of the following device.

Lemma 5.39. *Let K be a weakly compact subset of a separable Banach space X. Then there is a norm, $|||\cdot|||$, on X with the following properties:*

(a) $|||x||| \le \|x\|$ *for every* $x \in X$;
(b) K *is compact in the normed space* $(X, |||\cdot|||)$;
(c) *The weak topologies associated with the norms* $\|\cdot\|$ *and* $|||\cdot|||$ *agree on* K.

Proof. Since X is separable, the unit sphere of X^* contains a countable norming subset, $\{\varphi_n\}$. Let

$$|||x||| = \sum_{n=1}^{\infty} 2^{-n}|\varphi_n(x)|\,.$$

It is easy to see that this defines a norm on X and that $|||x||| \le \|x\|$ for every $x \in X$. Now the series defining $|||x|||$ converges uniformly on every bounded subset of X. Therefore, if a sequence (x_k) converges weakly in X to x, then $|||x_k - x||| \to 0$. It follows from the Eberlein–Smulian Theorem that the weakly compact set K is compact for the norm $|||\cdot|||$. Finally, since the $\|\cdot\|$-weak topology is finer than the $|||\cdot|||$-weak topology, and K is compact in the former and Hausdorff in the latter, it follows that these topologies coincide on K. □

We can now establish the representability of every weakly compact operator with separable range.

Proposition 5.40. *Let μ be a finite, positive measure and let X be a separable Banach space. Then every weakly compact operator from $L_1(\mu)$ into X is representable.*

Proof. Let $T\colon L_1(\mu) \to X$ be weakly compact and let K be the closure of the image under T of the closed unit ball of $L_1(\mu)$. Let Z be the completion of the normed space $(X, |||\cdot|||)$, where $|||\cdot|||$ is a norm on X with the properties described in the lemma. Since X is separable, it follows from part (a) of the lemma that Z is also separable. Let J denote the injection of X into Z. Then JT is a compact operator and so there exists a μ-measurable, μ-essentially bounded function $g\colon \Omega \to Z$, such that

$$JTf = \int_\Omega fg \, d\mu$$

for every $f \in L_1(\mu)$. Applying this representation to the unit norm function $f = \mu(E)^{-1}\chi_E$, where $\mu(E) > 0$, we obtain

$$\frac{1}{\mu(E)} \int_E g \, d\mu \in K$$

for every such E. It follows from this fact that g takes almost all its values in the set K. Indeed, by the separability of Z, it suffices to prove that for every open ball $B = B(z, r)$ in Z that is disjoint from K, the set $E = g^{-1}(B)$ has measure zero. Suppose this were false. Then

$$\left\| z - \frac{1}{\mu(E)} \int_E g \, d\mu \right\| = \frac{1}{\mu(E)} \left\| \int_E (z - g) \, d\mu \right\| \le \frac{1}{\mu(E)} \int_E \|z - g\| \, d\mu \le r \,,$$

a contradiction. Hence $\mu(E) = 0$ and our assertion is proved. Therefore, by the Pettis Measurability Theorem, g may be considered to be a μ-measurable function from Ω into X. Furthermore, g is μ-essentially bounded as a function with values in X and hence the Bochner integral $\int_\Omega fg \, d\mu$ exists in X for every $f \in L_1(\mu)$. Therefore T is representable. □

The technique we have employed here to convert a weakly compact set into a compact set can also be adapted to deal with weak*-compact sets in certain dual spaces. Suppose that X is the dual of a separable Banach space Y and that X itself is also separable. Taking $\{y_n\}$ to be a countable dense subset of the unit sphere of Y, we can define a norm on X by $|||x||| = \sum_n 2^{-n}|x(y_n)|$ as before and the statement of the lemma now applies to any weak*-compact subset K of X, with the use of the Eberlein–Smulian Theorem replaced by an appeal to the metrizability of the weak* topology on K. This yields the following result.

Proposition 5.41. *Let μ be a finite, positive measure and let X be a separable dual space. Then every operator from $L_1(\mu)$ into X is representable.*

We now have our first example of a large class of infinite dimensional spaces with the Radon–Nikodým property:

Corollary 5.42. *Every separable dual space has the Radon–Nikodým property.*

We have seen that every weakly compact operator on $L_1(\mu)$ with values in a separable Banach space is representable. We shall extend this result to all Banach spaces by proving the following remarkable fact: every weakly compact operator on $L_1(\mu)$ takes its values in a separable space. We begin with an observation about weak compactness in $L_1(\mu)$, where μ is a finite, positive measure. The canonical injection of $L_\infty(\mu)$ into $L_1(\mu)$ factors through the canonical injection of $L_p(\mu)$ into $L_1(\mu)$ for every p and hence is weakly compact. This enables us to identify an interesting class of weakly compact sets:

Proposition 5.43. *Let μ be a finite, positive measure. Then every bounded subset of $L_\infty(\mu)$ is relatively weakly compact in $L_1(\mu)$.*

In particular, we see that the set $\{\chi_E\}$ of characteristic functions of measurable sets is relatively weakly compact in $L_1(\mu)$.

Theorem 5.44. *Let μ be a finite, positive measure and let X be a Banach space. Then every weakly compact operator from $L_1(\mu)$ into X takes its values in a separable subspace of X and is, therefore, representable.*

Proof. Let $T\colon L_1(\mu) \to X$ be a weakly compact operator. We claim that the set $\{T\chi_E : E \in \Sigma\}$ is relatively compact in X. Since the characteristic functions span a dense subspace of $L_1(\mu)$ and every compact set is separable, it then follows that T maps $L_1(\mu)$ into a separable subset of X.

To establish this claim, let (E_n) be a sequence of measurable sets. Let Σ_0 be the σ-algebra generated by these sets and let μ_0 be the restriction of μ to Σ_0. Since Σ_0 is generated by a countable family of sets, the subspace $L_1(\mu_0)$ of $L_1(\mu)$ is separable. Let T_0 denote the restriction of T to this subspace. Then T_0 is weakly compact and takes its values in a separable subspace of X and so, by Proposition 5.40, T_0 is representable. By Corollary 5.37, T_0 is completely continuous and hence maps the relatively weakly compact set $\{\chi_{E_n}\}$ into a relatively compact subset of X. Thus $(T\chi_{E_n})$ is a relatively compact sequence in X and our claim is proved. □

The fact that weakly compact operators on $L_1(\mu)$ are representable has many applications. First, since every operator with values in a reflexive space is weakly compact, we have:

Corollary 5.45. *Every reflexive Banach space has the Radon–Nikodým property.*

Next, we recall that the Banach space X has the Dunford–Pettis property if, for every Banach space Y, every weakly compact operator from X into Y is completely continuous. Since every representable operator is completely continuous, we have:

Corollary 5.46. *The space $L_1(\mu)$ has the Dunford–Pettis property.*

The next result combines the representability of weakly compact operators with the Lewis–Stegall Theorem and the fact that the space ℓ_1 has the Radon–Nikodým property to yield a striking and unexpected consequence.

Theorem 5.47. *Let $S\colon X \to Y$ be an integral operator and $T\colon Y \to Z$ a weakly compact operator. Then the composition TS is nuclear.*

Proof. Since S is integral, there exist a finite, positive measure μ and operators $B \in \mathcal{L}(X, L_\infty(\mu))$ and $A \in \mathcal{L}(L_1(\mu), Y^{**})$ such that $JS = AIB$, where J is the canonical embedding of Y into Y^{**} and I is the canonical injection from $L_\infty(\mu)$ into $L_1(\mu)$. Now T^{**} is a weakly compact operator from Y^{**} into Z. Therefore, the composition $T^{**}A\colon L_1(\mu) \to Z$ is weakly compact and hence factors through ℓ_1. These factorizations are shown in the diagram below:

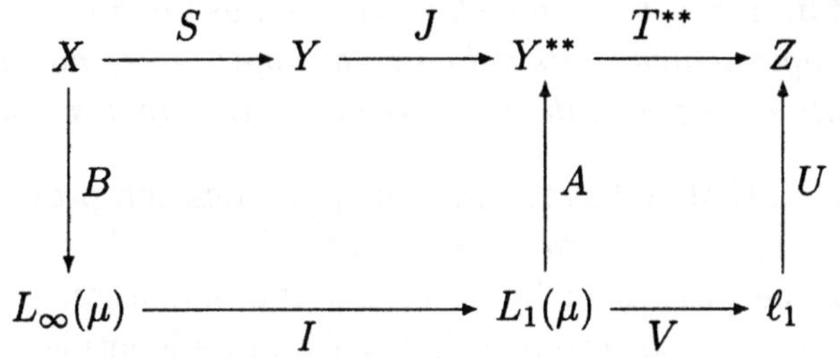

The operator I is Pietsch integral and hence so is VI. Since ℓ_1 has the Radon–Nikodým property, this operator is nuclear by Theorem 5.32 and hence $TS = T^{**}JS = U(VI)B$ is also nuclear. □

Next, we show that the Radon–Nikodým property is "separably determined".

Proposition 5.48. *If every separable subspace of a Banach space X has the Radon–Nikodým property, then X itself has the Radon–Nikodým property.*

Proof. Let T be an operator from $L_1(\mu)$ into X. We show that T takes its values in a separable subspace of X by proving that the set $\{T\chi_E : E \in \Sigma\}$ is relatively compact. Proceeding as in the proof of Theorem 5.44, we see that every sequence (E_n) of measurable sets lies in a countably generated sub-σ-algebra Σ_0 of Σ and the restriction of T to $L_1(\mu_0)$ then takes its values in a separable subspace of X. Therefore $(T\chi_{E_n})$ is a relatively compact sequence in X. □

The space $\ell_1(I)$, where I is an uncountable indexing set, is an example of a space that satisfies the hypothesis of this proposition. Since each element of $\ell_1(I)$ is supported by a countable subset of I, it follows that every separable subspace is contained in a subspace that is isometrically isomorphic to ℓ_1.

Proposition 5.49. *If the Banach space X has the Radon–Nikodým property, then so does every closed subspace of X.*

Proof. Let T be an operator from $L_1(\mu)$ into a closed subspace Y of X. Thus there exists a μ-measurable, μ-essentially bounded function g, with values in X, such that $Tf = \int_\Omega fg\,d\mu$ for every $f \in L_1(\mu)$. Let $Q\colon X \to X/Y$ be the quotient operator. Then

$$\int_\Omega f(Qg)\,d\mu = QTf = 0$$

for every $f \in L_1(\mu)$. The function Qg, being μ-measurable, is essentially separably valued and hence $Qg = 0$ almost everywhere. Therefore g is μ-essentially Y-valued and so T is representable. □

We finish this section with another striking application of the Radon–Nikodým property.

Theorem 5.50. *Let X be a Banach space that has the Radon–Nikodým property and is complemented in its bidual by a projection of norm one. If X has the approximation property, then X has the metric approximation property.*

Proof. We recall that X has the metric approximation property if and only if the canonical mapping from $X^*\hat{\otimes}_\pi X$ into $(X\hat{\otimes}_\varepsilon X^*)^*$ is an isometric embedding. Now the possession by X of the approximation property implies that the projective tensor product $X^*\hat{\otimes}_\pi X$ can be identified with the space $\mathcal{N}(X,X)$ of nuclear operators on X. Since X has the Radon–Nikodým property, the nuclear and Pietsch integral operators on X coincide and the nuclear norm is equal to the Pietsch integral norm. The condition that X is complemented in X^{**} by a projection of unit norm implies that the spaces of Pietsch integral and integral operators on X are the same. To summarize, we have

$$X^*\hat{\otimes}_\pi X = \mathcal{I}(X,X)\,.$$

On the other hand, the dual space $(X\hat{\otimes}_\varepsilon X^*)^*$ can be identified with the space $\mathcal{I}(X,X^{**})$. Therefore, with these identifications, the canonical mapping from $X^*\hat{\otimes}_\pi X$ into $(X\hat{\otimes}_\varepsilon X^*)^*$ is nothing more than the canonical embedding of $\mathcal{I}(X,X)$ into $\mathcal{I}(X,X^{**})$. □

Corollary 5.51. *If X is a reflexive space, or a separable dual space, and has the approximation property, then X has the metric approximation property.*

5.5 The Principle of Local Reflexivity

In this section, we prove a fundamental fact about the bidual of a Banach space. This result, known as the Principle of Local Reflexivity, shows that the finite dimensional, or "local", structure of X^{**} mimics that of X very closely. This principle is rooted in the fact that finite dimensional spaces have the Radon–Nikodým property. Our first proposition makes essential use of this observation.

Proposition 5.52. *Let E be a finite dimensional Banach space and let X be any Banach space. Then $(\mathcal{L}(E,X))^{**} = \mathcal{L}(E,X^{**})$.*

Proof. Every operator from E into X is approximable and hence $\mathcal{L}(E,X) = E^*\otimes_\varepsilon X$. Since E^* has both the Radon–Nikodým property and the approximation property, it follows from Theorem 5.33 that

$$\mathcal{L}(E,X)^* = (E^* \otimes_\varepsilon X)^* = E \otimes_\pi X^* .$$

Therefore, we have

$$\mathcal{L}(E,X)^{**} = (E \otimes_\pi X^*)^* = \mathcal{L}(E,X^{**}) .$$

□

As the dual space of $\mathcal{L}(E,X)$ is the tensor product $E \otimes_\varepsilon X^*$, the action of $T \in \mathcal{L}(E,X^{**})$ as an element of the bidual $\mathcal{L}(E,X)^{**}$ is completely determined by the equation

$$\langle x \otimes \varphi, T \rangle = \langle \varphi, Tx \rangle ,$$

where $x \in E$ and $\varphi \in X^*$.

We can say a little more when the finite dimensional space E is contained in X^{**}. First, we introduce some notation. If E is a subspace of X^{**}, we shall denote by $\mathcal{L}_0(E,X)$, the closed subspace of $\mathcal{L}(E,X)$ consisting of those operators that act as the identity on $E \cap X$. In other words, T belongs to $\mathcal{L}_0(E,X)$ if $Tx = x$ for every $x \in E \cap X$. Similarly, $\mathcal{L}_0(E,X^{**})$ will denote the subspace of $\mathcal{L}(E,X^{**})$ consisting of the operators acting as the identity on $E \cap X$.

Proposition 5.53. *Let X be a Banach space and let E be a finite dimensional subspace of X^{**}. Then $(\mathcal{L}_0(E,X))^{**} = \mathcal{L}_0(E,X^{**})$.*

Proof. We wish to show that $(\mathcal{L}_0(E,X))^{**}$ is the subspace $\mathcal{L}_0(E,X^{**})$ of $\mathcal{L}(E,X^{**})$. Suppose first that the operator $T\colon E \to X^{**}$ belongs to $(\mathcal{L}_0(E,X))^{**}$. Then, by Goldstine's Theorem, there is a bounded net, (T_α), in $\mathcal{L}_0(E,X)$ that converges in the weak* topology to T. Thus

$$\langle T_\alpha x, \varphi \rangle = \langle T_\alpha, x \otimes \varphi \rangle \to \langle x \otimes \varphi, T \rangle = \langle \varphi, Tx \rangle$$

for every $x \in E$ and $\varphi \in X^*$. Since $T_\alpha x = x$ for every $x \in E \cap X$ and every α, it follows that $Tx = x$ for every $x \in E \cap X$. Therefore $T \in \mathcal{L}_0(E,X^{**})$.

Conversely, suppose $T \in \mathcal{L}_0(E,X^{**})$. Let F be a finite dimensional subspace of E such that E is the direct sum of $E \cap X$ and F. Choosing a basis $\{f_1,\ldots,f_m\}$ for F, we can write each $x \in E$ as $u + \sum_{j=1}^m x_j f_j$, where $u \in E \cap X$. Therefore $Tx = u + \sum_{j=1}^m x_j T f_j$. Let $(z_{1\alpha})_\alpha, \ldots, (z_{m\alpha})_\alpha$ be nets in X that converge in the weak* topology of X^{**} to $Tf_1, \ldots, Tf_m$ respectively. Consider the net of operators (T_α), where

$$T_\alpha x = u + \sum_{j=1}^m x_j z_{j\alpha} .$$

Then each T_α belongs to $\mathcal{L}_0(E,X)$ and it is easy to see that the net (T_α) is weak* convergent to T in $\mathcal{L}(E,X)^{**}$. Therefore $T \in (\mathcal{L}_0(E,X))^{**}$. □

In the application of this proposition, we shall make use of Helly's Lemma, in the following form: let Z be a Banach space, z^{**} an element of Z^{**}, M a finite dimensional subspace of Z^* and $\varepsilon > 0$. Then there exists $z \in Z$ with $\|z\| < \|z^{**}\| + \varepsilon$, such that $\langle z, \varphi\rangle = \langle \varphi, z^{**}\rangle$ for every $\varphi \in M$.

Theorem 5.54 (The Principle of Local Reflexivity). *Let X be a Banach space and let E and F be finite dimensional subspaces of X^{**} and X^* respectively. Then, for every $\varepsilon > 0$, there exists an operator $T\colon E \to X$ with the following properties:*

(a) $(1-\varepsilon)\|x^{**}\| \le \|Tx^{**}\| \le (1+\varepsilon)\|x^{**}\|$ *for every* $x^{**} \in E$.
(b) $\langle Tx^{**}, \varphi\rangle = \langle \varphi, x^{**}\rangle$ *for every* $x^{**} \in E$, $\varphi \in F$.
(c) $Tx = x$ *for every* $x \in E \cap X$.

Proof. Since E is finite dimensional, we may assume, by enlarging the subspace F if necessary, that for every $x^{**} \in E$ there exists $\varphi_{x^{**}}$ in the unit sphere of F such that

$$(1-\varepsilon)\|x^{**}\| \le \langle \varphi_{x^{**}}, x^{**}\rangle .$$

We now apply Helly's lemma to the bidual of the Banach space $Z = \mathcal{L}_0(E, X)$, namely $\mathcal{L}_0(E, X^{**})$. Let z^{**} be the embedding of E into X^{**} and M the finite dimensional subspace of $\mathcal{L}(E, X)^*$ spanned by the functionals $x^{**} \otimes \varphi$, as x^{**} and φ range over E and F respectively. Therefore, there exists an operator T in $\mathcal{L}_0(E, X)$ such that $\|T\| < 1 + \varepsilon/2$ and

$$\langle Tx^{**}, \varphi\rangle = \langle \varphi, x^{**}\rangle \qquad \text{for every } x^{**} \in E,\ \varphi \in F, \tag{5.5}$$

and so T satisfies (b). Applying (5.5) with $\varphi = \varphi_{x^{**}}$, we obtain

$$(1-\varepsilon)\|x^{**}\| \le \|Tx^{**}\|$$

for every $x^{**} \in E$. Therefore, T satisfies (a). Finally, (c) is simply the definition of $\mathcal{L}_0(E, X)$. □

The Principle of Local Reflexivity has fundamental consequences and we shall make extensive use of it. Indeed, many of the results in the preceding chapters that relate to the bidual take on a new meaning in the light of this result. For now, we illustrate its power with a couple of applications.

We have already encountered the classes of $\mathcal{L}_1$-spaces and $\mathcal{L}_\infty$-spaces. It is a consequence of the principle of local reflexivity that X is an $\mathcal{L}_1$-space (respectively, an $\mathcal{L}_\infty$-space) if and only if the bidual X^{**} is an $\mathcal{L}_1$-space (respectively, an $\mathcal{L}_\infty$-space). We leave the details as an exercise for the reader.

For our second application, we prove a Schauder Theorem for approximable operators.

Proposition 5.55. *Let X and Y be Banach spaces. An operator $T\colon X \to Y$ is approximable if and only if the adjoint operator T^* is approximable.*

Proof. If T is approximable, it is easy to see that T^* is approximable. Suppose that T^* is approximable. Let J denote the canonical embedding of Y into Y^{**}. Then $JT\colon X \to Y^{**}$, being the restriction of the approximable operator T^{**} to X, is approximable. Let $\varepsilon > 0$. Then there exists a finite rank operator $S\colon X \to Y^{**}$ such that $\|JT - S\| < \varepsilon$. Now T is compact, since its adjoint is, and so we may choose an ε-net, $\{Tx_1, \ldots, Tx_n\}$ for the relatively compact subset TB_X of Y. Let E be the finite dimensional subspace of Y^{**} spanned by $\{Tx_1, \ldots, Tx_n\} \cup S(X)$. By the Principle of Local Reflexivity, there exists an operator $R\colon E \to Y$ such that $\|R\| \le 1 + \varepsilon$ and T is the identity on $E \cap Y$. We claim that the finite rank operator RS from X into Y approximates T. To see this, fix $x \in B_X$ and choose x_i so that $\|Tx - Tx_i\| < \varepsilon$. Since $Tx_i \in E \cap Y$, we have $RTx_i = Tx_i$ and hence

$$Tx - RSx = (Tx - Tx_i) + (RTx_i - RTx) + (RTx - RSx).$$

It follows that $\|T - RS\| \le (3 + 2\varepsilon)\varepsilon$. Therefore T is approximable. □

5.6 Exercises

Exercise 5.1. Show that the space of vector measures of bounded variation is complete in the variation norm.

Exercise 5.2. Show that the semivariation of a vector measure μ on a set E is given by

$$|\mu|_\infty(E) = \sup\left\{\left\|\sum_{i=1}^{n} \alpha_i \mu(A_i)\right\| : |\alpha_i| = 1, \{A_1, \ldots, A_n\} \text{ a partition of } E\right\}.$$

Exercise 5.3. Show that the semivariation of a vector measure is countably additive if and only if it coincides with the variation.

Exercise 5.4. Let Σ be the σ-algebra of Lebesgue measurable subsets of $[0,1]$ and let $\mu\colon \Sigma \to L_p[0,1]$ be the vector measure given by $\mu(E) = \chi_E$, where $1 \le p < \infty$. Show that μ has bounded variation if and only if $p = 1$.

Exercise 5.5. Find an example of a function $\mu\colon \Sigma \to X^*$ that is weak*-countably additive, but is not a vector measure (weak*-countable additivity means that the scalar function $E \mapsto \langle x, \mu(E)\rangle$ is a measure for every $x \in X$).

Exercise 5.6. Let $\mu\colon \Sigma \to X$ be a vector measure and $T\colon X \to Y$ an operator. Show that $T\mu$ is a vector measure that has bounded variation if μ has bounded variation, with $\|T\mu\|_1 \le \|T\|\|\mu\|_1$ and $\|T\mu\|_\infty \le \|T\|\|\mu\|_\infty$.

Exercise 5.7. Show that the space of measures $\mathcal{M}(\Sigma)$ has the metric approximation property.

Exercise 5.8. Show that $\mathcal{M}(\Sigma)$ is an $\mathcal{L}_{1,\lambda}$-space for every $\lambda > 1$.

Exercise 5.9. Show that a vector measure $\mu \colon \Sigma \to X$ has the Radon–Nikodým property if and only if the associated operator $T_\mu \colon X^* \to \mathcal{M}(\Sigma)$ is nuclear.

Exercise 5.10. Show that the ℓ_1-sum $(\sum_n X_n)_1$ of a sequence of Banach spaces with the Radon–Nikodým property also has this property. (The space $(\sum_n X_n)_1$ consists of all sequences $x = (x_n)$, where $x_n \in X_n$ for each n and $\|x\| := \sum_n \|x_n\| < \infty$.)

Exercise 5.11. Show that the space $C(K)^{**}$ is injective. (Hint: show that projective tensor products with $C(K)^*$ respect subspaces.)

Exercise 5.12. Show that the operator $T \colon L_1[0,1] \to C[0,1]$ defined by

$$Tf(t) = \int_0^t f(s)\,ds$$

is not representable.

Exercise 5.13. Find an operator from $L_1[0,1]$ into c_0 that is not representable.

Exercise 5.14. Let $S \colon X \to Y$ be a weakly compact operator and $T \colon Y \to Z$ an integral operator and suppose that X^* has the approximation property. Show that the composition TS is nuclear.

Exercise 5.15. Show that the composition of two integral operators is a nuclear operator.

Exercise 5.16. Use the principle of local reflexivity to show that $X \otimes_\pi Y$ is a subspace of $X^{**} \otimes_\pi Y^{**}$ for every pair of Banach spaces X, Y.

6. The Chevet–Saphar Tensor Products

In this chapter we begin a general study of tensor norms. Taking the projective and injective norms as our models, we formulate the appropriate definition of a tensor norm. We then investigate the Chevet–Saphar tensor norms, g_p and d_p. The dual spaces of the corresponding tensor products lead us to the definition of the p-summing operators. We conclude with the fundamental Grothendieck Inequality and some of its applications.

6.1 Tensor Norms

What properties should a tensor product norm possess? In this section, we will attempt to provide a satisfactory answer to this question, taking the projective and injective norms as models.

Let X and Y be Banach spaces. We say that a norm, α, on $X \otimes Y$ is a *reasonable crossnorm* if it has the following properties:

1. $\alpha(x \otimes y) \leq \|x\|\|y\|$ for every $x \in X$ and $y \in Y$.
2. For every $\varphi \in X^*$ and $\psi \in Y^*$, the linear functional $\varphi \otimes \psi$ on $X \otimes Y$ is bounded, and $\|\varphi \otimes \psi\| \leq \|\varphi\|\|\psi\|$.

Of course, the projective and injective norms satisfy these conditions. Our next result shows that these norms lie at opposite ends of the spectrum of reasonable crossnorms.

Proposition 6.1. *Let X and Y be Banach spaces.*

(a) *A norm α on $X \otimes Y$ is a reasonable crossnorm if and only if*

$$\varepsilon(u) \leq \alpha(u) \leq \pi(u)$$

for every $u \in X \otimes Y$.

(b) *If α is a reasonable crossnorm on $X \otimes Y$ then $\alpha(x \otimes y) = \|x\|\|y\|$ for every $x \in X$ and $y \in Y$. Furthermore, for every $\varphi \in X^*$ and $\psi \in Y^*$, the norm of the linear functional $\varphi \otimes \psi$ on $(X \otimes Y, \alpha)$ satisfies $\|\varphi \otimes \psi\| = \|\varphi\|\|\psi\|$.*

Proof. (a) Suppose that α is a reasonable crossnorm on $X \otimes Y$. Then, for every representation $\sum_{j=1}^{n} x_j \otimes y_j$ of $u \in X \otimes Y$,

$$\alpha(u) \le \sum_{j=1}^n \alpha(x_j \otimes y_j) \le \sum_{j=1}^n \|x_j\|\|y_j\|$$

and it follows that $\alpha(u) \le \pi(u)$. Also, since $\|\varphi \otimes \psi\| \le 1$ whenever $\varphi \in B_{X^*}$, $\psi \in B_{Y^*}$, we have

$$\begin{aligned}\varepsilon(u) &= \sup\{|\langle u, \varphi \otimes \psi\rangle : \varphi \in B_{X^*}, \psi \in B_{Y^*}\} \\ &\le \sup\{|\langle u, v\rangle : v \in (X \otimes Y, \alpha)^*,\ \|v\| \le 1\} = \alpha(u)\,.\end{aligned}$$

Conversely, suppose that α is a norm on $X \otimes Y$ that lies between the injective and projective norms. Then $\alpha(x \otimes y) \le \pi(x \otimes y) = \|x\|\|y\|$ for every $x \in X$ and $y \in Y$. Furthermore, since $\alpha(u) \le 1$ implies $\varepsilon(u) \le 1$,

$$\|\varphi \otimes \psi\| \le \sup\{|\langle u, \varphi \otimes \psi\rangle| : \varepsilon(u) \le 1\} = \|\varphi\|\|\psi\|$$

for every $\varphi \in X^*$ and $\psi \in Y^*$. Therefore α is a reasonable crossnorm.

(b) If α is a reasonable crossnorm on $X \otimes Y$, then $\alpha(x \otimes y) = \|x\|\|y\|$ follows immediately from the fact that $\varepsilon(x \otimes y) \le \alpha(x \otimes y) \le \pi(x \otimes y)$. Let $\varphi \in X^*$ and $\psi \in Y^*$. Then, using the fact that $\pi \ge \alpha$,

$$\|\varphi \otimes \psi\| \ge \sup\{|\langle u, \varphi \otimes \psi\rangle| : \pi(u) \le 1\} = \|\varphi\|\|\psi\|$$

and it follows that $\|\varphi \otimes \psi\| = \|\varphi\|\|\psi\|$. □

We have seen that the injective and projective tensor products behave well with respect to the formation of tensor product of operators: if $S\colon X \to W$ and $T\colon Y \to Z$ are operators, then the operator $S \otimes T\colon X \otimes Y \to W \otimes Z$ is bounded and $\|S \otimes T\| \le \|S\|\|T\|$. This is clearly a desirable attribute and so we add it to our requirements for a tensor product norm. A *uniform crossnorm* is an assignment to each pair, X, Y of Banach spaces of a reasonable crossnorm on $X \otimes Y$ for which the above property holds.

If α is a uniform crossnorm and $u \in X \otimes Y$, we shall denote the α-norm of u by $\alpha(u; X \otimes Y)$ or $\alpha_{X,Y}(u)$, or simply $\alpha(u)$ if there is no risk of ambiguity. The tensor product $X \otimes Y$ with this norm will be denoted by $X \otimes_\alpha Y$ and its completion by $X \hat{\otimes}_\alpha Y$. If $S\colon X \to W$ and $T\colon Y \to Z$ are operators, then the operator $S \otimes T\colon X \otimes_\alpha Y \to W \otimes_\alpha Z$ satisfies $\|S \otimes T\| \le \|S\|\|T\|$. It has a unique extension to an operator from $X \hat{\otimes}_\alpha Y$ into $W \hat{\otimes}_\alpha Z$ with the same norm. We shall denote this operator by $S \otimes_\alpha T$.

If α and β are uniform crossnorms, we shall interpret relations such as $\alpha \le \beta$ to mean that $\alpha(u; X \otimes Y) \le \beta(u; X \otimes Y)$ for every pair of Banach spaces X, Y and every $u \in X \otimes Y$.

We make some elementary observations about uniform crossnorms. First, we consider the norm induced on the tensor product of subspaces. Let E, F be subspaces of X, Y respectively. Then the tensor product $E \otimes F$ is an algebraic subspace of $X \otimes Y$. Applying the uniform property to the tensor product of the inclusion operators $E \to X$ and $F \to Y$, we see that

$$\alpha_{X,Y}(u) \leq \alpha_{E,F}(u) \tag{6.1}$$

for every $u \in E \otimes F$. From our experience of the projective norm, we know that this inequality can be strict. Thus $E \otimes_\alpha F$ need not be a subspace of $X \otimes_\alpha Y$. On the other hand, we have seen that the injective norm does respect subspaces. We say that a uniform crossnorm is *injective* if it shares this property. Thus, α is injective if, whenever E, F are subspaces of X, Y respectively, then the norm induced on $E \otimes F$ by the norm of $X \otimes_\alpha Y$ coincides with the norm of $E \otimes_\alpha F$.

Next, we consider quotient spaces. Let $Q\colon X \to W$ and $R\colon Y \to Z$ be quotient operators. Then, by the uniform property, the tensor product operator $Q \otimes_\alpha R\colon X \otimes_\alpha Y \to W \otimes_\alpha Z$ has unit norm. We shall say that the uniform crossnorm α is *projective* if this operator is always a quotient operator. We have seen that the projective norm has this property: if W, Z are quotients of X, Y respectively, then $W \otimes_\pi Z$ is a quotient of $X \otimes_\pi Y$.

There is one property of the injective and projective norms that we do not generalize. These norms are symmetric – interchanging the factor spaces does not alter the norm. If α is a uniform crossnorm, the *transpose* of α is the uniform crossnorm α^t defined as follows: for each pair of Banach spaces X, Y and each $u \in X \otimes Y$,

$$\alpha^t(u; X \otimes Y) = \alpha(u^t; Y \otimes X),$$

where the transpose, u^t, of $u = \sum_{j=1}^n x_j \otimes y_j$ is given by $u^t = \sum_{j=1}^n y_j \otimes x_j$. We say that α is *symmetric* if $\alpha^t = \alpha$. We shall encounter many examples of non-symmetric norms.

There is just one more ingredient needed for a satisfactory theory of tensor product norms. We recall that, although the projective norm is not injective, the projective norm of a tensor $u \in X \otimes Y$ is determined by the values of the projective norms of u in all the finite dimensional tensor products $M \otimes N$ that contain u. We have made essential use of this fact in our discussion of $\mathcal{L}_1$-spaces and $\mathcal{L}_\infty$-spaces in previous chapters. And the Principle of Local Reflexivity demonstrates in a dramatic way the importance of the finite dimensional structure of a Banach space.

We say that a uniform crossnorm α is *finitely generated* if, for every pair X, Y of Banach spaces and every $u \in X \otimes Y$, we have

$$\alpha(u; X \otimes Y) = \inf\{\alpha(u; M \otimes N) : u \in M \otimes N,\ \dim M,\ \dim N < \infty\}. \tag{6.2}$$

Thus, the behaviour of α is completely determined by its values on tensor products of finite dimensional spaces. Occasionally, it will be useful to say that a uniform crossnorm is finitely generated on a certain class of Banach spaces (or even on a single pair of spaces) if (6.2) holds for every pair of spaces X, Y belonging to the given class.

The reader may find it difficult to find an example of a uniform crossnorm that is not finitely generated. In the light of our next result, this is not at all surprising.

Proposition 6.2. *Let X and Y be Banach spaces with the metric approximation property. Then every uniform crossnorm is finitely generated on $X \otimes Y$.*

Proof. Let α be a uniform crossnorm. It follows from (6.1) that

$$\alpha(u; X \otimes Y) \leq \inf\{\alpha(u; M \otimes N) : u \in M \otimes N, \quad \dim M,\ \dim N < \infty\}$$

for every $u \in X \otimes Y$. To prove the reverse inequality, let $u \in X \otimes Y$ and let $\varepsilon > 0$. Fix a representation $\sum_{j=1}^{n} x_j \otimes y_j$ of u. Using the metric approximation property, there exist finite rank operators S, T on X, Y respectively, such that $\|S\|$, $\|T\| \leq 1$ and $\|Sx_j - x_j\|$, $\|Ty_j - y_j\| < \varepsilon'$ for $1 \leq j \leq n$, where ε' is chosed so that $(\sum_{j=1}^{n} \|x_j\| + \|y_j\|)\varepsilon' < \varepsilon$. Now choose finite dimensional subspaces M, N of X, Y respectively so that M contains $\{x_1, \ldots, x_n\}$ and the range of S, while N contains $\{y_1, \ldots, y_n\}$ and the range of T. Then both u and $S \otimes T(u)$ belong to $M \otimes N$ and

$$\begin{aligned}
\alpha_{M,N}(u - S \otimes T(u)) &= \alpha_{M,N}\left(\sum_{j=1}^{n} (x_j \otimes y_j - (Sx_j) \otimes (Ty_j))\right) \\
&= \alpha_{M,N}\left(\sum_{j=1}^{n} ((x_j - Sx_j) \otimes y_j + (Sx_j) \otimes (y_j - Ty_j))\right) \\
&\leq \sum_{j=1}^{n} (\|x_j - Sx_j\| \|y_j\| + \|Sx_j\| \|y_j - Ty_j\|) < \varepsilon \,.
\end{aligned}$$

If we consider S, T as operators from X, Y into M, N respectively, then, by the uniform property of α, we have

$$\alpha_{M,N}(S \otimes T(u)) \leq \|S\| \|T\| \, \alpha_{X,Y}(u) \leq \alpha_{X,Y}(u) \,.$$

Putting these inequalities together, we obtain

$$\alpha_{M,N}(u) < \alpha_{M,N}(u - S \otimes T(u)) + \alpha_{M,N}(S \otimes T(u)) < \alpha_{X,Y}(u) + \varepsilon \,.$$

We have shown that, for every $\varepsilon > 0$ there exist finite dimensional subspaces M, N of X, Y respectively, such that $\alpha_{M,N}(u) < \alpha_{X,Y}(u) + \varepsilon$. Therefore $\alpha_{X,Y}(u) = \inf \alpha_{M,N}(u)$ and the proof is complete. □

We define a *tensor norm* to be a finitely generated uniform crossnorm. At this point, we have only two examples of tensor norms – the projective and the injective norm. We note in passing that the property of finite generation is trivial for ε and indeed for every injective uniform crossnorm.

Our next result gives two simple applications of finite generation. The first illustrates how finite dimensional information can be used to infer general facts and the second shows how to construct a tensor norm from a uniform crossnorm on finite dimensional spaces. The proofs are straightforward and are left to the reader.

Proposition 6.3.

(a) *Let α, β be tensor norms. Then $\alpha \le \beta$ if and only if $\alpha_{E,F} \le \beta_{E,F}$ for every pair E, F of finite dimensional normed spaces.*
(b) *Let γ be a uniform crossnorm defined on the class of finite dimensional normed spaces. Then the formula*

$$\gamma(u; X \otimes Y) = \inf\{\gamma(u; M \otimes N) : u \in M \otimes N, \quad \dim M,\ \dim N < \infty\}$$

defines a tensor norm that coincides with γ on tensor products of finite dimensional normed spaces.

There is an important special case in which every tensor norm behaves in an injective fashion:

Proposition 6.4. *Let α be a tensor norm and let X, Y be Banach spaces. Then $X \otimes_\alpha Y$ is a subspace of $X^{**} \otimes_\alpha Y^{**}$.*

Proof. By the uniform property of α, we have $\alpha_{X^{**},Y^{**}} \le \alpha_{X,Y}$ on $X \otimes Y$. To prove the reverse inequality, let $u \in X \otimes Y$ and let $\varepsilon > 0$. Fix a representation $\sum_{j=1}^n x_j \otimes y_j$ of u. Since α is finitely generated, there exist finite dimensional subspaces M, N of X^{**}, Y^{**} respectively such that

$$\alpha_{M,N}(u) \le \alpha_{X^{**},Y^{**}}(u) + \varepsilon\,.$$

Now the value of $\alpha_{M,N}(u)$ does not increase if M and N are enlarged. Hence we may assume that $\{x_1, \ldots, x_n\} \subset M$ and $\{y_1, \ldots, y_n\} \subset N$. By the Principle of Local Reflexivity, there exist finite dimensional subspaces E, F of X, Y respectively and isomorphisms $S\colon M \to E$ and $T\colon N \to F$, each of norm at most $1 + \varepsilon$, such that $Sx_j = x_j$ and $Ty_j = y_j$ for each j. It follows that $S \otimes T(u) = u$ and hence

$$\begin{aligned}\alpha_{X,Y}(u) &\le \alpha_{E,F}(u) = \alpha_{E,F}(S \otimes T(u)) \\ &\le (1+\varepsilon)^2 \alpha_{M,N}(u) < (1+\varepsilon)^2 \alpha_{X^{**},Y^{**}}(u) + \varepsilon(1+\varepsilon)^2\,.\end{aligned}$$

Since $\varepsilon > 0$ is arbitrary, we have $\alpha_{X,Y}(u) \le \alpha_{X^{**},Y^{**}}(u)$. □

We conclude this section with some remarks about the dual of a tensor product. Let α be a reasonable crossnorm on $X \otimes Y$. Since α is dominated by the projective norm, every bounded linear functional on the Banach space $X \hat{\otimes}_\alpha Y$ is given by a unique bounded bilinear form on $X \times Y$. On the other hand, the injective norm is the smallest reasonable crossnorm, and so every integral bilinear form on $X \times Y$ is also bounded for the norm α. Thus, the dual space of $X\hat{\otimes}_\alpha Y$ is a space of bilinear forms that lies somewhere between these extremes:

$$\mathcal{B}_I(X \times Y) \subset (X\hat{\otimes}_\alpha Y)^* \subset \mathcal{B}(X \times Y).$$

Of course, the elements of this dual space may also be viewed as operators, either from X into Y^* or from Y into X^*.

We recall from Chapter 2 that every bounded bilinear form on $X \times Y$ has an extension to a bounded bilinear form on $X^{**} \times Y^{**}$ with the same norm. In fact, there are two "natural" ways to construct this extension. If $B \in \mathcal{B}(X \times Y)$ corresponds to the operator $T : X \to Y^*$, so that $B(x, y) = \langle y, Tx\rangle$, we may define

$$^*B(x^{**}, y^{**}) = \langle T^* y^{**}, x^{**}\rangle .$$

We shall refer to this as the *canonical left extension* of B. On the other hand, we have $B(x, y) = \langle x, Sy\rangle$, where $S \in \mathcal{L}(Y, X^*)$. Thus we may define

$$B^*(x^{**}, y^{**}) = \langle S^* x^{**}, y^{**}\rangle .$$

We call B^* the *canonical right extension* of B.

Theorem 6.5. *Let α be a tensor norm, let X and Y be Banach spaces and let $B \in (X\hat{\otimes}_\alpha Y)^*$. Then the canonical left and right extensions of B are bounded linear functionals on $X^{**}\hat{\otimes}_\alpha Y^{**}$ with the same norm as B.*

Proof. We give the proof for the canonical left extension. Let $T\colon X \to Y^*$ be the operator associated to B. Let $u \in X^{**} \otimes Y^{**}$ and choose a representation $\sum_{j=1}^n x_j^{**} \otimes y_j^{**}$ for u. Then $\langle u, {}^*B\rangle = \sum_{j=1}^n \langle T^* y_j^{**}, x_j^{**}\rangle$. We apply the Principle of Local Reflexivity to X^{**}; take M_1 and N_1 to be the subspaces of X^{**} and X^* spanned by $\{x_1^{**}, \ldots, x_n^{**}\}$ and $\{x_1^*, \ldots, x_n^*\}$ respectively and take $\varepsilon > 0$. There exists an subspace E_1 of X and an isomorphism $R\colon M_1 \to E_1$ of norm at most $1 + \varepsilon$ such that $\langle Rx^{**}, x^*\rangle = \langle x^*, x^{**}\rangle$ for every $x^{**} \in M_1$ and $x^* \in N_1$. Therefore,

$$\langle u, {}^*B\rangle = \sum_{j=1}^n \langle T^* y_j^{**}, x_j^{**}\rangle = \sum_{j=1}^n \langle Rx_j^{**}, T^* y_j^{**}\rangle = \sum_{j=1}^n \langle TRx_j^{**}, y_j^{**}\rangle .$$

We apply the Principle of Local Reflexivity now to Y^{**}, taking M_2 and N_2 to be the spaces spanned by the y_j^{**} and y_j^* respectively. This yields an isomorphism, S, from M_2 into a subspace E_2 of Y of norm at most $1 + \varepsilon$, and we have

$$\sum_{j=1}^n \langle TRx_j^{**}, y_j^{**}\rangle = \sum_{j=1}^n \langle Sy_j^{**}, TRx_j^{**}\rangle = \left\langle \sum_{j=1}^n Rx_j^{**} \otimes Sy_j^{**}, B\right\rangle .$$

Now, by the uniform property of α, the tensor product operator $R\otimes S\colon M_1\otimes_\alpha M_2 \to E_1 \otimes_\alpha E_2$ has norm at most $(1+\varepsilon)^2$. Thus,

$$|\langle u, {}^*B\rangle| \le \alpha_{X,Y}\left(\sum_{j=1}^n Rx_j^{**}\otimes Sy_j^{**}\right)\|B\| \le \alpha_{E_1,E_2}\left(\sum_{j=1}^n Rx_j^{**}\otimes Sy_j^{**}\right)\|B\|$$
$$\le (1+\varepsilon)^2\alpha_{M_1,M_2}\left(\sum_{j=1}^n x_j^{**}\otimes y_j^{**}\right)\|B\|\,.$$

Since α is finitely generated, we may choose M_1 and M_2 so that

$$\alpha_{M_1,M_2}\left(\sum_{j=1}^n x_j^{**}\otimes y_j^{**}\right)$$

is arbitrarily close to $\alpha_{X^{**},Y^{**}}(\sum_{j=1}^n x_j^{**}\otimes y_j^{**})$ and it follows that $\|{}^*B\| = \|B\|$. □

6.2 The Chevet–Saphar Tensor Norms

We now construct some new tensor norms. Let us begin with another look at the definition of the injective norm on the tensor product $X\otimes Y$. We recall that this norm may be defined by embedding the tensor product in the space of operators $\mathcal{L}(X^*,Y)$. Thus, if $\sum_{j=1}^n x_j\otimes y_j$ is any representation of the tensor $u\in X\otimes Y$, then

$$\varepsilon(u) = \sup\left\{\left\|\sum_{j=1}^n \varphi(x_j)y_j\right\| : \varphi\in B_{X^*}\right\}.$$

Since ε is the smallest tensor norm, we look for a norm that dominates this expression. If we fix $p\in(1,\infty)$ with conjugate index p', then applying Hölder's Inequality yields

$$\varepsilon(u) \le \sup_{\varphi\in B_{X^*}}\left(\sum_{j=1}^n |\varphi(x_j)|^{p'}\right)^{1/p'}\left(\sum_{j=1}^n \|y_j\|^p\right)^{1/p}.$$

A brief calculation shows that this expression is dominated by the projective norm, $\pi(u)$, and so we are on the right track. However, the formula above now depends on the representation chosen for u. Borrowing an idea from the definition of the projective norm, we arrive at the following candidate for a tensor norm:

$$\|u\| = \inf\left\{\sup_{\varphi\in B_{X^*}}\left(\sum_{j=1}^n |\varphi(x_j)|^{p'}\right)^{1/p'}\left(\sum_{j=1}^n \|y_j\|^p\right)^{1/p}\right\}, \tag{6.3}$$

the infimum being taken over the set of all representations of the tensor u. This formula is rather cumbersome and it will be helpful if we pause to organize our notation.

For $1 \le p < \infty$, we shall denote by $\ell_p(X)$ the space of all p-summable sequences in the Banach space X, that is, sequences (x_n) with the property that the series $\sum_{n=1}^{\infty} \|x_n\|^p$ converges. A norm is defined on this space by

$$\|(x_n)\|_p = \left(\sum_{n=1}^{\infty} \|x_n\|^p\right)^{1/p}.$$

The obvious modification is made in the case $p = \infty$, so that $\ell_\infty(X)$ is simply the space of bounded sequences in X with the norm $\|(x_n)\|_\infty = \sup_n \|x_n\|$. We leave it to the reader to verify that $\ell_p(X)$ is a Banach space.

There are also weak versions of these spaces. We denote by $\ell_p^w(X)$ the space of weakly p-summable sequences in X, that is, sequences (x_n) with the property that the scalar sequence (φx_n) belongs to ℓ_p for every $\varphi \in X^*$. An application of the Closed Graph Theorem shows that we may associate with each weakly p-summable sequence (x_n) an operator $S\colon X^* \to \ell_p$, where $S\varphi = (\varphi x_n)$. Now the adjoint operator sends the unit vectors e_n to the points x_n. It follows that S^* maps $\ell_{p'}$ into X when $1 < p < \infty$ and when $p = 1$, S^* maps c_0 into X. We define a norm on the space $\ell_p^w(X)$ by taking the norm of the operator S or the norm of its adjoint:

$$\begin{aligned} \|(x_n)\|_p^w &= \sup\left\{\left(\sum_{n=1}^{\infty} |\varphi x_n|^p\right)^{1/p} : \varphi \in B_{X^*}\right\} \\ &= \sup\left\{\left\|\sum_{n=1}^{\infty} \lambda_n x_n\right\|_X : \lambda \in B_{\ell_{p'}}\right\} \qquad \text{when } 1 < p < \infty, \end{aligned} \tag{6.4}$$

and

$$\begin{aligned} \|(x_n)\|_1^w &= \sup\left\{\sum_{n=1}^{\infty} |\varphi x_n| : \varphi \in B_{X^*}\right\} \\ &= \sup\left\{\left\|\sum_{n=1}^{\infty} \lambda_n x_n\right\|_X : \lambda \in B_{c_0}\right\} \qquad \text{when } p = 1. \end{aligned}$$

In applying these formulas, we may replace the closed unit balls B_{X^*}, $B_{\ell_{p'}}$ or B_{c_0} by norming sets.

A moment's thought reveals that $\ell_p^w(X)$ may be identified with the space $\mathcal{L}(\ell_{p'}, X)$ when $1 < p < \infty$, or with $\mathcal{L}(c_0, X)$ when $p = 1$. It follows that these sequence spaces are complete. Since the weakly bounded and norm bounded sequences are the same, the space $\ell_\infty^w(X)$ coincides with $\ell_\infty(X)$ and the weak and strong norms are the same in this case.

We return now to our putative tensor norm. We may write (6.3) using the compact notation developed above, with the understanding that every finite sequence $(x_1, \ldots, x_n)$ may be interpreted as the infinite sequence $(x_1, \ldots, x_n, 0, 0, \ldots)$ when it appears in a formula for an ℓ_p-norm. The *Chevet–Saphar norms* are defined for $1 \le p \le \infty$ as follows:

$$d_p(u) = \inf\left\{ \|(x_j)\|_{p'}^w \, \|(y_j)\|_p : u = \sum_{j=1}^n x_j \otimes y_j \right\}, \tag{6.5}$$

taking the infimum over all representations of $u \in X \otimes Y$. We may interchange the roles of the weak and strong norms to define a parallel sequence of norms:

$$g_p(u) = \inf\left\{ \|(x_j)\|_p \, \|(y_j)\|_{p'}^w : u = \sum_{j=1}^n x_j \otimes y_j \right\}. \tag{6.6}$$

It may be helpful to the reader to know that the notation for these norms comes from the French words "gauche" and "droite". Clearly, we have

$$g_p = d_p^t$$

for every p. It follows that every property of d_p can be interpreted as a property of g_p and vice versa.

We shall find the following variations on (6.5) useful:

$$d_p(u) = \inf\left\{ \|(\lambda_j)\|_p \, \|(x_j)\|_{p'}^w \, \|(y_j)\|_\infty : u = \sum_{j=1}^n \lambda_j x_j \otimes y_j \right\}. \tag{6.7}$$

Of course, there is a similar variation on the formula defining g_p. To prove (6.7), let us denote the expression given there by $\delta_p(u)$ for the moment. We note first that every representation of u of the form $\sum_{j=1}^n \lambda_j x_j \otimes y_j$ can be written as $\sum_{j=1}^n x_j \otimes (\lambda_j y_j)$ and hence $d_p(u) \le \|(x_j)\|_{p'}^w \, \|(\lambda_j y_j)\|_p \le \|(\lambda_j)\|_p \, \|(x_j)\|_{p'}^w \, \|(y_j)\|_\infty$, from which it follows that $d_p(u) \le \delta_p(u)$. On the other hand, let $\sum_{j=1}^n x_j \otimes y_j$ be a representation of u. We can write u as $\sum_{j=1}^n \lambda_j x_j \otimes z_j$, where $\lambda_j = \|y_j\|$ and $\|z_j\| \le 1$ for every j. Then $\delta_p(u) \le \|(y_j)\|_p \, \|(x_j)\|_{p'}^w$ and hence $\delta_p(u) \le d_p(u)$.

Proposition 6.6. *Let* $1 \le p, q \le \infty$.

(a) d_p *and* g_p *are tensor norms.*
(b) *If* $p \le q$ *then* $d_p \ge d_q$ *and* $g_p \ge g_q$.
(c) $d_1 = g_1 = \pi$.

Proof. (a) We show first that d_p is a reasonable crossnorm on every tensor product $X \otimes Y$. We deal with the case $1 < p < \infty$ and we leave it to the reader to make the appropriate small modifications in the other cases. Let u_1, $u_2 \in X \otimes Y$ and let $\varepsilon > 0$. Choose representations $u_i = \sum_{j=1}^n x_{ij} \otimes y_{ij}$ such that $\|(x_{ij})_j\|_{p'}^w \le (d_p(u_i) + \varepsilon)^{1/p}$ and $\|(y_{ij})_j\|_p \le (d_p(u_i) + \varepsilon)^{1/p'}$ for $i = 1, 2$. Then $\sum_{i,j} x_{ij} \otimes y_{ij}$ is a representation of $u_1 + u_2$ with

$$\|(x_{ij})\|_{p'}^{w}\,\|(y_{ij})\|_p \le \left(d_p(u_1)+d_p(u_2)+2\varepsilon\right)^{1/p}\left(d_p(u_1)+d_p(u_2)+2\varepsilon\right)^{1/p'}$$
$$= d_p(u_1)+d_p(u_2)+2\varepsilon$$

and it follows that $d_p(u_1+u_2) \le d_p(u_1)+d_p(u_2)$. It is easy to see that $d_p(\lambda u) = |\lambda| d_p(u)$ for every $\lambda \in \mathbb{K}$ and $u \in X \otimes Y$. Therefore d_p is a seminorm. But it is clear that $\varepsilon(u) \le d_p(u) \le \pi(u)$ for every u, and hence d_p is a reasonable crossnorm.

The uniform property for d_p follows easily from the definition. Finally, the fact that the weak and strong ℓ_p-norms are unchanged if the range space is enlarged shows that d_p is finitely generated. Therefore d_p is a tensor norm.

(b) Suppose that $1 < p < q < \infty$. Let $u \in X \otimes Y$ and let $\varepsilon > 0$. It will suffice to show that $d_q(u) \le d_p(u)+\varepsilon$. By (6.7), we may choose a representation of u of the form $\sum_{j=1}^n \lambda_j x_j \otimes y_j$ such that

$$\|(\lambda_j)\|_p\,\|(x_j)\|_{p'}^{w}\,\|(y_j)\|_\infty \le d_p(u)+\varepsilon\,.$$

We rewrite this representation as

$$u = \sum_{j=1}^n \lambda_j^{p/q}(\lambda_j^{1-p/q}x_j)\otimes y_j$$

so that

$$d_q(u) \le \|(\lambda_j^{p/q})\|_q\,\|(\lambda_j^{1-p/q}x_j)\|_{q'}^{w}\,\|(y_j)\|_\infty\,. \tag{6.8}$$

Consider the weak $\ell_{q'}$-norm

$$\|(\lambda_j^{1-p/q}x_j)\|_{q'}^{w} = \sup_{\varphi\in B_{X^*}}\left(\sum_{j=1}^n |\lambda_j|^{(1-p/q)q'}|\varphi x_j|^{q'}\right)^{1/q'}.$$

Fixing φ, we apply Hölder's Inequality with the conjugate exponents $p'/(p'-q')$ and p'/q', to get

$$\left(\sum_{j=1}^n |\lambda_j|^{(1-p/q)q'}|\varphi x_j|^{q'}\right)^{1/q'} \le \left(\sum_{j=1}^n |\lambda_j|^p\right)^{1/p-1/q}\left(\sum_{j=1}^n |\varphi x_j|^{p'}\right)^{1/p'}.$$

Taking the supremum over $\varphi \in B_{X^*}$ and using (6.8), we have

$$d_q(u) \le \|(\lambda_j)\|_p\,\|(x_j)\|_{p'}^{w}\,\|(y_j)\|_\infty \le d_p(u)+\varepsilon$$

and we are done. We leave it to the reader to dispose of the case $q=\infty$. The case $p=1$ will follow from (c).

(c) If $\sum_{j=1}^n x_j\otimes y_j$ is a representation of u, then $\pi(u) \le \sum_{j=1}^n \|x_j\|\|y_j\| \le \|(x_j)\|_\infty\,\|(y_j)\|_1$. Therefore $\pi(u) \le d_1(u)$. The reverse inequality follows from (a). □

We say that a tensor norm α is *right projective* if, for every Banach space X and every quotient operator $Q\colon Z \to Y$, the tensor product operator $I \otimes_\alpha Q\colon X\hat{\otimes}_\alpha Z \to X\hat{\otimes}_\alpha Y$ is a quotient operator. There is a similar definition for *left projective* tensor norms.

Proposition 6.7. *The tensor norm d_p is right projective for every p.*

Proof. Let $Q\colon Z \to Y$ be a quotient operator. We show that the operator $I \otimes Q\colon X \otimes_\alpha Z \to X \otimes_\alpha Y$ is a quotient operator for every X. We have $\|I \otimes Q\| = 1$ and so it suffices to prove that $I \otimes Q$ maps the open unit ball of $X \otimes_\alpha Z$ onto the open unit ball of $X \otimes_\alpha Y$. Let $u \in X \otimes_\alpha Y$ with $d_p(u) < 1$. Choose $\varepsilon > 0$ so that $(1+\varepsilon)d_p(u) < 1$. Now let $\sum_{j=1}^n x_j \otimes y_j$ be a representation of u such that $\|(x_j)\|_{p'}^w \, \|(y_j)\|_p < 1$. Since Q is a quotient operator, we may choose $z_j \in Z$ such that $Qz_j = y_j$ and $\|z_j\| \leq (1+\varepsilon)\|y_j\|$ for every j. Let $v = \sum_{j=1}^n x_j \otimes z_j \in X \otimes Z$. Then $I \otimes Q(v) = u$ and $d_p(v) \leq (1+\varepsilon)\|(x_j)\|_{p'}^w \, \|(y_j)\|_p < 1$. □

Of course, there is a corresponding result for the norms g_p – these norms are all left projective. When $p = 1$, then $d_1 = g_1 = \pi$ are both left and right projective.

Bearing in mind the coincidence of the projective norm on $\ell_1 \otimes X$ with the norm of $\ell_1(X)$, it is natural to ask if there is an analogous description of the Chevet–Saphar norms on $\ell_p \otimes X$. However, when $p > 1$, there are two norms, d_p and g_p, in play and the outcome is not quite as pleasing:

Example 6.8. *The Chevet–Saphar norms g_p and d_p on $\ell_p \otimes X$.*

Let $1 < p < \infty$. We may embed $\ell_p \otimes X$ algebraically as a subspace of $\ell_p(X)$ in the usual way. If $u = \sum_{j=1}^n a_j \otimes x_j \in \ell_p \otimes X$, where $a_j = (a_{jm})_m$, we may write u as $\sum_{m=1}^\infty e_m \otimes u_m$, where $u_m = \sum_{j=1}^n a_{jm}x_j$. Thus, u is identified with the element (u_m) of $\ell_p(X)$. Now the unit basis vectors e_m in ℓ_p satisfy $\|(e_m)\|_{p'}^w = 1$ and so

$$d_p(u) \leq \|u\|_p\,.$$

On the other hand, with the help of Hölder's Inequality, we have

$$\begin{aligned}
\|u\|_p &= \left(\sum_{m=1}^\infty \left\|\sum_{j=1}^n a_{jm}x_j\right\|^p\right)^{1/p} = \left(\sum_{m=1}^\infty \left(\sup_{\varphi\in B_{X^*}} \left|\sum_{j=1}^n a_{jm}\varphi(x_j)\right|\right)^p\right)^{1/p} \\
&\leq \left(\sum_{m=1}^\infty \left(\left(\sum_{j=1}^n |a_{jm}|^p\right)^{1/p} \sup_{\varphi\in B_{X^*}} \left(\sum_{j=1}^n |\varphi(x_j)|^{p'}\right)^{1/p'}\right)^p\right)^{1/p} \\
&= \left(\sum_{m=1}^\infty \sum_{j=1}^n |a_{jm}|^p\right)^{1/p} \sup_{\varphi\in B_{X^*}} \left(\sum_{j=1}^n |\varphi(x_j)|^{p'}\right)^{1/p'} \\
&= \left(\sum_{j=1}^n \|(a_{jm})_m\|^p\right)^{1/p} \sup_{\varphi\in B_{X^*}} \left(\sum_{j=1}^n |\varphi(x_j)|^{p'}\right)^{1/p'} = \|(a_j)\|_p \, \|(x_j)\|_{p'}^w\,.
\end{aligned}$$

This holds for every representation of u and so, by the definition of g_p,

$$\|u\|_p \le g_p(u)\,.$$

Putting these results together, we see that

$$d_p(u) \le \|u\|_p \le g_p(u) \tag{6.9}$$

for every $u \in \ell_p \otimes X$. In general, this is the best that can be said about these norms. We refer to the exercises for some related results.

We turn our attention now to deriving a description of the elements of the completed tensor product $X\hat{\otimes}_{d_p}Y$ similar to that proved in Chapter 2 for the projective tensor product.

Lemma 6.9. *Let X and Y be Banach spaces.*

(a) *Let $1 \le p < \infty$. If $(x_n) \in \ell_{p'}^w(X)$ and $(y_n) \in \ell_p(Y)$, then the series $\sum_{n=1}^\infty x_n \otimes y_n$ converges in $X\hat{\otimes}_{d_p}Y$.*
(b) *If $(x_n) \in \ell_1^w(X)$ and the sequence (y_n) converges to zero in Y, then the series $\sum_{n=1}^\infty x_n \otimes y_n$ converges in $X\hat{\otimes}_{d_\infty}Y$.*

Proof. We prove (a); the proof of (b) is virtually the same and is left to the reader. Let us denote by $(x_j)_n^m$ the finite sequence $(x_n, \ldots, x_m)$, where $n \le m$. We have

$$d_p\Big(\sum_{j=n}^m x_j \otimes y_j\Big) \le \ell_{p'}^w((x_j)_n^m)\,\ell_p((y_j)_n^m) \le \ell_{p'}^w((x_j))\Big(\sum_{j=n}^m \|y_j\|^p\Big)^{1/p}$$

for $n \le m$. It follows that the series $\sum_{j=1}^\infty x_j \otimes y_j$ satisfies the Cauchy condition. □

We show now that every element of $X\hat{\otimes}_{d_p}Y$ is the sum of a series of this form.

Proposition 6.10. *Let $1 \le p \le \infty$ and let X, Y be Banach spaces. If $u \in X\hat{\otimes}_{d_p}Y$ and $\varepsilon > 0$, then there exist sequences $(x_n) \in \ell_{p'}^w(X)$ and $(y_n) \in \ell_p(Y)$ (or $c_0(Y)$ when $p = \infty$) such that the series $\sum_{n=1}^\infty x_n \otimes y_n$ converges to u and*

$$d_p(u) \le \|(x_n)\|_{p'}^w\, \|(y_n)\|_p \le d_p(u) + \varepsilon\,.$$

Proof. The result has already been established for the projective norm (Proposition 2.8) and so we may assume that $1 < p \le \infty$. Let $\eta > 0$. Since $X \otimes Y$ is a dense subspace of $X\hat{\otimes}_{d_p}Y$, we may apply a standard argument to obtain a sequence (u_m) in $X \otimes Y$ such that $u = \sum_{m=1}^\infty u_m$ with $d_p(u_1) < d_p(u) + \eta$ and $d_p(u_m) < \eta^2/4^m$ when $m \ge 2$.

We choose a representation of u_1 of the form $\sum_{j=1}^{k_1} x_{1j} \otimes y_{1j}$ such that

$$\|(x_{1j})_j\|_{p'}^w < d_p(u) + \eta \qquad \text{and} \qquad \|(y_{1j})_j\|_p \leq 1\,.$$

For each $m \geq 2$, choose a representation of u_m of the form $\sum_{j=1}^{k_m} x_{mj} \otimes y_{mj}$ such that

$$\|(x_{mj})_j\|_{p'}^w < \eta/2^m \qquad \text{and} \qquad \|(y_{mj})_j\|_p < \eta/2^m\,.$$

Concatenating the finite sequences $(x_{mj})_j$ and $(y_{mj})_j$, we obtain infinite sequences (x_n) and (y_n). Thus, there is a strictly increasing sequence of positive integers $1 = l_1 < l_2 < \dots$, so that the blocks in the sequences (x_n) and (y_n) corresponding to $l_m \leq n < l_{m+1}$ coincide with the finite sequences $(x_{mj})_j$ and $(y_{mj})_j$ respectively.

We claim that the series $\sum_{n=1}^{\infty} x_n \otimes y_n$ converges to u and has the required properties. We note first that

$$\|(y_n)\|_p = \left(\sum_{m=1}^{\infty}\sum_{j=1}^{k_m} \|y_{mj}\|^p\right)^{1/p} < \left(1 + \eta^p \sum_{m=2}^{\infty} 2^{-pm}\right)^{1/p},$$

for $1 < p < \infty$. When $p = \infty$, we have $(y_n) \in c_0(Y)$ and we may assume that η is small enough so that $\sup \|y_n\| \leq 1$.

To compute the weak $\ell_{p'}$-norm of (x_n), fix $\varphi \in B_{X^*}$. Then

$$\left(\sum_{n=1}^{\infty} |\varphi x_n|^{p'}\right)^{1/p'} = \left(\sum_{m=1}^{\infty}\sum_{j=1}^{k_m} |\varphi x_{mj}|^{p'}\right)^{1/p'} \leq \left(\sum_{m=1}^{\infty}\left(\|(x_{mj})_j\|_{p'}^w\right)^{p'}\right)^{1/p'}.$$

Taking the supremum over φ, we obtain

$$\|(x_n)\|_{p'}^w \leq \left((d_p(u) + \eta)^{p'} + \eta^{p'} \sum_{m=2}^{\infty} 2^{-p'm}\right)^{1/p'}$$

It follows from the Lemma that the series $\sum_{n=1}^{\infty} x_n \otimes y_n$ converges in $X \hat{\otimes}_{d_p} Y$ and it is clear that its sum is u. It is easy to see that $d_p(u) \leq \|(x_n)\|_{p'}^w \, \|(y_n)\|_p$. Finally, using the estimates we have obtained for $\|(x_n)\|_{p'}^w$ and $\|(y_n)\|_p$, if η is sufficiently small, then $\|(x_n)\|_{p'}^w \, \|(y_n)\|_p \leq d_p(u) + \varepsilon$. □

We now consider how the Chevet–Saphar tensor products can be interpreted as spaces of operators. We recall that the elements of the projective tensor product $X^* \hat{\otimes}_\pi Y$ can be viewed as nuclear operators from X into Y. In the same way, there is a canonical linear mapping from $X^* \otimes_{g_p} Y$ into $\mathcal{L}(X, Y)$ for each p that associates with the tensor $\sum_{j=1}^{n} \varphi_j \otimes y_j$ the operator $x \mapsto \sum_{j=1}^{n} \varphi_j(x) y_j$. An application of Hölder's Inequality shows that this mapping is bounded. Therefore it extends to an operator $J_p\colon X^* \hat{\otimes}_{g_p} Y \to \mathcal{L}(X, Y)$. An operator from X into Y is called *p-nuclear* (or sometimes *left p-nuclear*) if it lies in the range of J_p. Thus, T is a p-nuclear operator if there exist sequences $(\varphi_n) \in \ell_p(X^*)$ (or $c_0(X^*)$ when $p = \infty$) and $(y_n) \in \ell_{p'}^w(Y)$, such that $Tx = \sum_{n=1}^{\infty} \varphi_n(x) y_n$ for every $x \in X$. We denote by $\mathcal{N}_p(X,Y)$ the space of p-nuclear operators and we define the *p-nuclear norm* on this space by

$$\|T\|_{N_p} = \inf \|(\varphi_n)\|_p \, \|(y_n)\|_{p'}^{w} ,$$

where the infimum is taken over all the representations of T as described above. We leave it to the reader to verify that $\mathcal{N}_p(X,Y)$ is complete in this norm. Replacing the norm g_p by d_p in this construction leads to the class of *right p-nuclear operators.* Since $g_1 = d_1 = \pi$, the (left) 1-nuclear and right 1-nuclear operators coincide with the nuclear operators.

It is clear from these definitions that the space of p-nuclear operators is a quotient of the tensor product $X^*\hat{\otimes}_{g_p}Y$:

$$\mathcal{N}_p(X,Y) = X^*\hat{\otimes}_{g_p}Y / \ker J_p \, .$$

We shall see later that $\mathcal{N}_p(X,Y) = X^*\hat{\otimes}_{g_p}Y$ if either X^* or Y has the approximation property.

6.3 p-summing Operators

The dual space of a tensor product $X\hat{\otimes}_\alpha Y$ can be interpreted as a space of bilinear forms on $X \times Y$, a space of operators from X into Y^*, or a space of operators from Y into X^*. A judicious choice often leads to the discovery of an exciting new type of bilinear form or operator. In this section, we shall study the dual of $X\hat{\otimes}_{d_p}Y$ as a space of operators from X into Y^*. The operators obtained in this way are known as the p'-summing operators.

Let $T \in \mathcal{L}(X,Y^*)$ be a bounded linear functional on $X\hat{\otimes}_{d_p}Y$. We may assume that $1 < p \leq \infty$, since d_1 is the projective norm. The action of T on a tensor $u = \sum_{j=1}^{n} x_j \otimes y_j$ is given by $\langle u, T\rangle = \sum_{j=1}^{n} \langle y_j, Tx_j\rangle$ and the boundedness of T means that

$$\left|\sum_{j=1}^{n} \langle y_j, Tx_j\rangle\right| \leq C \, \|(x_j)\|_{p'}^{w} \, \|(y_j)\|_p \, , \tag{6.10}$$

where C is a positive constant, the minimum value of which is the norm of T in $(X\hat{\otimes}_{d_p}Y)^*$. The condition given in (6.10) can be simplified as follows.

Proposition 6.11. *Let $1 < p \leq \infty$ and let X and Y be Banach spaces. An operator $T\colon X \to Y^*$ defines a bounded linear functional on $X\hat{\otimes}_{d_p}Y$ if and only if there exists a positive constant, C, such that*

$$\sum_{j=1}^{n} \|Tx_j\|^{p'} \leq C^{p'} \sup_{\varphi \in B_{X^*}} \sum_{j=1}^{n} |\varphi x_j|^{p'}$$

for every finite sequence $(x_1, \ldots, x_n)$ in X. Furthermore, the norm of T in $(X\hat{\otimes}_{d_p}Y)^*$ is the smallest value of C for which this condition holds.

Proof. Suppose that T satisfies this condition. Let $u = \sum_{j=1}^{n} x_j \otimes y_j \in X \otimes Y$. Then

$$|\langle u, T\rangle| = \left|\sum_{j=1}^{n} \langle y_j, Tx_j\rangle\right| \le \sum_{j=1}^{n} \|y_j\|\|x_j\| \le \left(\sum_{j=1}^{n} \|Tx_j\|^{p'}\right)^{1/p'} \left(\sum_{j=1}^{n} \|y_j\|^{p}\right)^{1/p}$$

by Hölder's Inequality, and it follows that T is bounded on $X \otimes_{d_p} Y$.

Conversely, suppose $T \in (X\hat{\otimes}_{d_p} Y)^*$, so that T satisfies (6.10) for some $C > 0$. Let $x_1, \ldots, x_n \in X$ and let $\varepsilon > 0$. For each j, choose $y_j \in Y$ such that $\langle y_j, Tx_j\rangle = \|Tx_j\|^{p'}$ and $\|y_j\| \le (1+\varepsilon)\|Tx_j\|^{p'-1}$. Then

$$\sum_{j=1}^{n} \|Tx_j\|^{p'} = \sum_{j=1}^{n} \langle y_j, Tx_j\rangle \le C(1+\varepsilon)\|(x_j)\|_{p'}^{w} \left(\sum_{j=1}^{n} \|Tx_j\|^{p'}\right)^{1/p}$$

since $p(p'-1) = p'$. It follows that $\sum_{j=1}^{n} \|Tx_j\|^{p'} \le C^{p'}(1+\varepsilon)^{p'}\|(x_j)\|_{p'}^{w}$ for every $\varepsilon > 0$. Therefore T and C satisfy the stated condition.

The assertion about the norm of T follows immediately from the proof.

□

At this point, it will be helpful if we interchange the conjugate indices p and p'. Thus, we have the following characterization of $T \in (X\hat{\otimes}_{d_{p'}} Y)^*$ for $1 \le p < \infty$:

$$\sum_{j=1}^{n} \|Tx_j\|^{p} \le C^p \sup_{\varphi \in B_{X^*}} \sum_{j=1}^{n} |\varphi x_j|^p . \tag{6.11}$$

Now consider the action of T on infinite sequences in X. Suppose that the sequence (x_n) is weakly p-summable. Then, from (6.11), we have

$$\sum_{n=1}^{k} \|Tx_n\|^p \le C^p \sup_{\varphi \in B_{X^*}} \sum_{n=1}^{k} |\varphi x_n|^p \le C^p\left(\|(x_n)\|_p^w\right)^p$$

for every k. It follows that the sequence (Tx_n) is strongly p-summable in Y^*, and that

$$\|(Tx_n)\|_p \le C\|(x_n)\|_p^w . \tag{6.12}$$

Conversely, if the operator $T\colon X \to Y^*$ sends weakly p-summable sequences in X to strongly p-summable sequences in Y^* and (6.12) is satisfied, then of course, T belongs to $(X\hat{\otimes}_{d_{p'}} Y)^*$. We have uncovered a new class of operators. The next proposition gives some equivalent descriptions of this class. In order to define these operators, it is not necessary that the range space be a dual.

Proposition 6.12. *Let $1 \le p < \infty$ and let X and Y be Banach spaces. The following are equivalent for an operator $T\colon X \to Y$:*

(i) *There exists a positive constant, C, such that*

$$\left|\sum_{j=1}^{n}\langle Tx_j, \psi_j\rangle\right| \le C\|(x_j)\|_p^w \,\|(\psi_j)\|_{p'}$$

for all finite sequences $(x_1, \dots, x_n)$ and $(\psi_1, \dots, \psi_n)$ of elements of X and Y^* respectively.

(ii) *There exists a positive constant, C, such that*

$$\sum_{j=1}^{n}\|(Tx_j)\|^p \le C^p \sup_{\varphi\in B_{X^*}} \sum_{j=1}^{n}|\varphi x_j|^p$$

for every finite sequence $(x_1, \dots, x_n)$ in X.

(iii) *For every weakly p-summable sequence (x_n) in X, the sequence (Tx_n) is strongly p-summable in Y.*

(iv) *There exists a positive constant, C, such that*

$$\|(Tx_n)\|_p \le C\|(x_n)\|_p^w$$

for every weakly p-summable sequence (x_n) in X.

Furthermore, the minimum value of the constant C is the same in each case.

Proof. The discussion above shows that (i), (ii) and (iv) are equivalent and that each implies (iii). We prove that (iii) implies (iv).

If T satisfies (iii), then we may define a mapping $S\colon \ell_p^w(X) \to \ell_p(Y)$ by $S((x_n)) = (Tx_n)$. It is easy to see that S is linear and it follows from the Closed Graph Theorem that S is bounded. Hence T satisfies (iv). □

An operator $T\colon X \to Y$ is said to be *p-summing* (or *absolutely p-summing*) if it satisfies the equivalent conditions of this proposition. The space of p-summing operators will be denoted by $\mathcal{P}_p(X,Y)$. The *p-summing norm* of $T \in \mathcal{P}_p(X,Y)$, denoted by $\pi_p(T)$, is defined to be the minimum value of the constant C that appears in the equivalent statements above. It is easy to see that $\|T\| \le \pi_p(T)$ for every p. The 1-summing operators are often referred to simply as *summing operators* (or *absolutely summing operators.*) We leave it to the reader to prove, with the help of the above proposition, that the space of p-summing operators is complete in the p-summing norm.

Returning to our discussion of the dual space of the Chevet–Saphar tensor products, we can now see that

$$(X\hat{\otimes}_{d_p}Y)^* = \mathcal{P}_{p'}(X, Y^*)$$

where $1 < p \le \infty$. And, of course, there is a corresponding statement for the transposed norm g_p:

$$(X\hat{\otimes}_{g_p}Y)^* = \mathcal{P}_{p'}(Y, X^*)\,.$$

In the latter case, the action of a p-summing operator $S\colon Y \to X^*$ on the tensor product is given by $\langle \sum_{j=1}^n x_j \otimes y_j, S\rangle = \sum_{j=1}^n \langle x_j, Sy_j\rangle$.

The space $\mathcal{P}_p(X,Y)$ is not necessarily a dual space. However, it follows immediately from Proposition 6.12 that the p-summing property of an operator, and the value of its p-summing norm, are unchanged if the range space is enlarged. Thus, $\mathcal{P}_p(X,Y)$ may be embedded in $\mathcal{P}_p(X,Y^{**})$ and we have

$$\mathcal{P}_p(X,Y) \subset \mathcal{P}_p(X,Y^{**}) = (X\hat{\otimes}_{d_{p'}}Y^*)^*,$$

where $1 \le p < \infty$. Thus, the elements of $\mathcal{P}_p(X,Y)$ act, in a canonical way, as bounded linear functionals on $X\hat{\otimes}_{d_{p'}}Y^*$; we have

$$\left\langle \sum_{j=1}^n x_j \otimes \psi_j, T \right\rangle = \sum_{j=1}^n \langle Tx_j, \psi_j\rangle .$$

We now collect together some of the elementary properties of p-summing operators.

Proposition 6.13.

(a) *Let $S \in \mathcal{P}_p(X,Y)$ and let $T\colon W \to X$ and $R\colon Y \to Z$ be operators. Then the composition RST is p-summing and*

$$\pi_p(RST) \le \|R\|\pi_p(S)\|T\| .$$

(b) *Let $T \in \mathcal{P}_p(X,Y)$. Then T^{**} is also p-summing and has the same p-summing norm as T.*

(c) *Let $1 \le p \le q < \infty$. Then $\mathcal{P}_p(X,Y) \subset \mathcal{P}_q(X,Y)$ and $\pi_q(T) \le \pi_p(T)$ for every $T \in \mathcal{P}_p(X,Y)$.*

Proof. (a) By multiplying R and T by appropriate constants, we may assume that $\|T\| = 1$, so that $T^*(B_{X^*}) \subset B_{W^*}$. Let $w_1, \ldots, w_n \in W$. Then

$$\begin{aligned}\sum_{j=1}^n \|RSTw_j\|^p \le \|R\|\sum_{j=1}^n \|STw_j\|^p &\le \|R\|\pi_p(S)^p \sup_{\varphi\in B_{X^*}} \sum_{j=1}^n |\varphi Tw_j|^p \\ &= \|R\|\pi_p(S)^p \sup_{\varphi\in B_{X^*}} \sum_{j=1}^n |(T^*\varphi)w_j|^p \\ &\le \|R\|\pi_p(S)^p \sup_{\psi\in B_{W^*}} \sum_{j=1}^n |\psi w_j|^p ,\end{aligned}$$

and it follows from Proposition 6.12 that RST is p-summing and that $\|RST\| \le \|R\|\pi_p(S)\|T\|$.

(b) By the remarks preceding the proposition, we may consider T as an element of $\mathcal{P}_p(X,Y^{**}) = (X\hat{\otimes}_{d_{p'}}Y^*)^*$. Then T^{**} is the canonical extension of T to a bounded linear functional on $X^{**}\hat{\otimes}_{d_{p'}}Y^{***}$ and the result follows from Theorem 6.5.

(c) follows immediately from the fact that if $p < q$ then $p' > q'$ and so $d_{p'} \leq d_{q'}$ on $X \otimes Y^{**}$. □

The fact that the norms d_p are finitely generated translates into a dual statement for the class of p-summing operators. We have seen that an operator $T\colon X \to Y$ is p-summing if and only if T defines a bounded linear functional on the tensor product $X \otimes_{d_{p'}} Y^*$. The finite generation of $d_{p'}$ means that

$$d_{p'}(u; X \otimes Y^*) = \inf\{d_{p'}(u; M \otimes N) : u \in M \otimes N\}$$

the infimum being taken over all pairs M, N of finite dimensional subspaces of X, Y^* respectively for which $u \in M \otimes N$. The finite dimensional subspaces of Y^* are in one to one correspondence with the subspaces of Y of finite codimension. Thus, arguing in exactly the same way as we did for integral operators in Section 3.5, we obtain the following finite dimensional characterization:

Proposition 6.14. *Let $1 \leq p < \infty$ and let X and Y be Banach spaces. An operator $T\colon X \to Y$ is p-summing if and only if there exists a positive constant C such that for every finite dimensional subspace E of X and every finite codimensional subspace F of Y, the finite dimensional operator*

$$Q_F T I_E\colon E \to X \to Y \to Y/F$$

satisfies $\pi_p(Q_F T I_E) \leq C$. Furthermore, we have $\pi_p(T) = \inf C$, where the infimum is taken over all such pairs E, F.

We now look at some examples of p-summing operators.

Example 6.15. *The canonical p-summing operator $C(K) \to L_p(\mu)$.*

Let μ be a regular, positive Borel measure on the compact space K and let $J\colon C(K) \to L_p(\mu)$ be the canonical mapping that associates with each continuous function its equivalence class in $L_p(\mu)$. Suppose that the sequence (f_n) is weakly p-summable in $C(K)$. Then the series $\sum_{n=1}^{\infty} |f_n(t)|^p$ converges for every $t \in K$ and it follows that the sum of this series is a continuous function on K. Then, by the Lebesgue Monotone Convergence Theorem,

$$\sum_{n=1}^{\infty} \|Jf_n\|^p = \sum_{n=1}^{\infty} \int_K |f_n(t)|^p \, d\mu(t) < \infty .$$

Therefore J is p-summing. We invite the reader to compute the p-summing norm of J.

There is also a measure theoretic version of this example:

Example 6.16. *The canonical injection $L_\infty(\mu) \to L_p(\mu)$ is p-summing.*

Let μ be a finite, positive measure on a σ-algebra of subsets of a set Ω and let $J_p\colon L_\infty(\mu) \to L_p(\mu)$ be the canonical injection. Let $f_1, \ldots, f_n \in L_\infty(\mu)$. By (6.4),

$$\|(f_j)\|_p^w = \sup\left\{ \left\| \sum_{j=1}^n \lambda_j f_j \right\|_\infty^p : (\lambda_j) \in B_{\ell_{p'}} \right\}.$$

It follows that, for every $\lambda_1, \ldots, \lambda_n$, we have

$$\left| \sum_{j=1}^n \lambda_j f_j(\omega) \right| \le \|\lambda\|_{p'} \|(f_j)\|_p^w \quad \mu\text{-almost everywhere.}$$

Therefore, there is a μ-null set E so that this is valid for all rational values of λ_j and all $\omega \notin E$ and hence it holds for all λ_j and all $\omega \notin E$. It follows that

$$\left(\sum_{j=1}^n |f_j(\omega)|^p \right)^{1/p} \le \|(f_j)\|_p^w$$

for almost every ω. Integrating, we obtain

$$\sum_{j=1}^n \|J_p f_j\|_p^p = \sum_{j=1}^n \int |f_j|^p \, d\mu \le \mu(\Omega) \left(\|(f_j)\|_p^w \right)^p.$$

Therefore J_p is p-summing and $\pi_p(J_p) \le \mu(\Omega)^{1/p}$. But $\mu(\Omega)^{1/p} = \|J_p\| \le \pi_p(J_p)$ and so $\pi_p(J_p) = \mu(\Omega)^{1/p}$.

Example 6.17. *The canonical injection $\ell_1 \to \ell_2$ is 1-summing.*

Let $x_1, \ldots, x_n \in \ell_1$, where $x_j = (x_{jk})_k$ for each j. Let r_k be the Rademacher functions. Applying the Khinchine Inequality in $L_1 = L_1[0,1]$, we have

$$\begin{aligned} \sum_{j=1}^n \|x_j\|_2 = \sum_{j=1}^n \left(\sum_{k=1}^\infty |x_{jk}|^2 \right)^{1/2} &\le A_1^{-1} \sum_{j=1}^n \left\| \sum_{k=1}^\infty x_{jk} r_k \right\|_{L_1} \\ &= A_1^{-1} \int_0^1 \sum_{j=1}^n \left| \sum_{k=1}^\infty x_{jk} r_k(t) \right| dt. \end{aligned}$$

For each t, the sequence $(r_k(t))$ is a unit vector in ℓ_∞ and so

$$\sum_{j=1}^n \left| \sum_{k=1}^\infty x_{jk} r_k(t) \right| \le \sup_{\varphi \in B_{\ell_\infty}} \sum_{j=1}^n |\varphi x_j| = \|(x_j)\|_1^w.$$

Therefore, $\sum_{j=1}^n \|x_j\|_2 \le A_1^{-1} \|(x_j)\|_1^w$.

In fact, we shall see shortly that every operator from ℓ_1 into ℓ_2 is 1-summing.

We now present a powerful characterization of the class of p-summing operators. This result will enable us to prove some of the deeper properties of these operators.

Theorem 6.18 (Pietsch Domination Theorem). *Let $1 \le p < \infty$, let X and Y be Banach spaces and let K be a weak*-closed norming subset of B_{X^*}, endowed with the weak* topology. An operator $T\colon X \to Y$ is p-summing if and only if there exists a regular Borel probability measure, μ, on K and a positive constant, C, such that*

$$\|Tx\|^p \le C^p \int_K |\varphi x|^p \, d\mu(\varphi)$$

for every $x \in X$. Furthermore, the p-summing norm of T is the minimum value of C.

Proof. Suppose that T satisfies the above estimate. Then

$$\sum_{j=1}^n \|x_j\|^p \le C^p \int_K \sum_{j=1}^n |\varphi x_j|^p \, d\mu(\varphi) \le C^p \big(\|(x_j)\|_p^w\big)^p,$$

for all $x_1, \ldots, x_n \in X$, and it follows that T is p-summing with $\pi_p(T) \le C$.

Conversely, suppose that T is a p-summing operator. For each finite subset A of X, consider the function f_A on K given by

$$f_A(\varphi) = \sum_{x \in A} \|Tx\|^p - \pi_p(T)^p \sum_{x \in A} |\varphi x|^p .$$

Then f_A is weak* continuous and the p-summing property of T implies that $f_A(\varphi) \le 0$ for at least one $\varphi \in K$. Let L be the subset of $C(K)$ consisting of all the functions f_A, as A ranges over the collection of finite subsets of X. It is easy to see that $\lambda f_A = f_{A'}$ for all positive λ, where $A' = \{\lambda^{1/p} x : x \in A\}$, and that $f_A + f_B = f_{A \cup B}$ if A and B are disjoint. It follows that L is a convex subset of $C(K)$.

Let $P = \{f \in C(K) : f(\varphi) > 0 \text{ for every } \varphi \in K\}$. Then P is a convex, open subset of $C(K)$ and is disjoint from L. Therefore, by the Hahn–Banach Theorem, there exists a regular Borel measure μ and a constant γ, such that $\langle f_A, \mu \rangle \le \gamma < \langle f, \mu \rangle$ for every $f_A \in L$ and $f \in P$. Since the zero function belongs to L, we have $\gamma \ge 0$ and so μ is a positive measure. Now, by multiplying γ and μ by appropriate constants, we may assume that μ is a probability measure. We now have $0 \le \gamma \le \langle f, \mu \rangle$ for every positive function f. But the value of $\langle f, \mu \rangle$ can be arbitrarily small and hence γ must be zero. Thus, $\langle f_A, \mu \rangle \le 0$ for every $f_A \in L$. Now, if we take A to be a singleton set $\{x\}$, we have $f_A(\varphi) = \|Tx\|^p - \pi_p(T)^p |\varphi x|^p$ and, bearing in mind that $\mu(K) = 1$, we obtain

$$\|Tx\|^p \leq \pi_p(T)^p \int_K |\varphi x|^p \, d\mu \, .$$

Finally, it is clear from the proof that $\pi_p(T)$ is the minimum value of C. □

It is possible to interpret the Pietsch Domination Theorem as a factorization scheme. Let $T\colon X \to Y$ be a p-summing operator and let K, μ and C be as above. We shall attempt to factor T through the canonical operator $J_p\colon C(K) \to L_p(\mu)$. Let V denote the embedding of X into $C(K)$, so that $(Vx)(\varphi) = \varphi(x)$ for $x \in X$ and $\varphi \in K$. Applying J_p takes us into the subspace $J_pV(X)$ of $L_p(\mu)$. We can now define $U\colon J_pV(X) \to Y$ by $U(J_pVx) = Tx$ and the domination condition tells us that U is bounded, since $\|U(J_pVx)\| \leq C\|J_pVx\|_p$ for every $x \in X$. Furthermore, we see that we choose K and μ so that $\|V\| \leq \pi_p(T)$. This factorization is shown below:

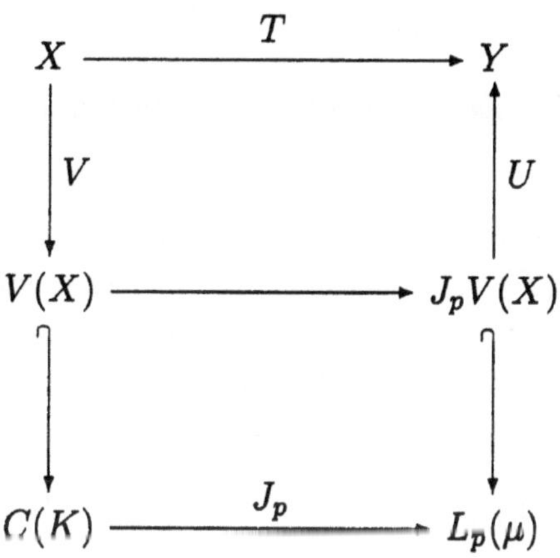

As it stands, this is not very satisfactory. Ideally, we would like to factor T through the canonical operator J_p. Thus, we need to extend $U\colon J_pV(X) \to Y$ to $L_p(\mu)$. In general, this will not be possible unless Y is an injective space. However, we may embed any Banach space Y as a subspace of an injective space. For example, let $\ell_\infty(B_{Y^*})$ be the ℓ_∞-space with indexing set B_{Y^*} (or any weak*-dense subset of this ball). Then Y may be embedded in this space by identifying $y \in Y$ with $(\psi y) \in \ell_\infty(B_{Y^*})$. Alternatively, we may compose the canonical embedding of Y into $C(B_{Y^*})$ with the embedding of the latter space into its bidual to obtain an embedding of Y into the injective space $C(B_{Y^*})^{**}$. We can now extend the operator U to $L_p(\mu)$ while preserving norm, giving a more attractive diagram:

$$\begin{array}{ccccc} X & \xrightarrow{T} & Y & \longrightarrow & \ell_\infty(B_{Y^*}) \\ \Big\downarrow V & & & & \Big\uparrow U \\ C(K) & & \xrightarrow{J_p} & & L_p(\mu) \end{array}$$

In addition, we have $\|V\|\|J_p\|\|U\| = \pi_p(T)$. We now show that the p-summing operators are characterized by factorizations of this general form.

Theorem 6.19. *Let $1 \leq p < \infty$ and let X and Y be Banach spaces. An operator $T\colon X \to Y$ is p-summing if and only if there exists a compact space K and a regular Borel probability measure μ on K, together with operators $V\colon X \to C(K)$ and $U\colon L_p(\mu) \to \ell_\infty(B_{Y^*})$, such that $IT = UJ_pV$, where I is the canonical embedding of Y into $\ell_\infty(B_{Y^*})$. Furthermore,*

$$\pi_p(T) = \inf \|U\|\|V\|,$$

where this infimum is taken over all such factorizations, and this infimum is attained

Proof. We have already seen that every p-summing operator can be factored in this way, with $\pi_p(T) = \inf \|U\|\|V\|$. Conversely, suppose that T can be factored in the manner described. Since J_p is p-summing, it follows from Proposition 6.13 that IT is p-summing. Therefore T is p-summing and has the same p-summing norm as T. Also,

$$\pi_p(T) = \pi_p(IT) = \pi_p(UJ_pV) \leq \|U\|\pi_p(J_p)\|V\| = \|U\|\|V\|.$$

This concludes the proof. □

To give an example of the power of this factorization, we recall from Chapter 3 that the operator $J_p\colon C(K) \to L_p(\mu)$ is both weakly compact and completely continuous and hence so also is the composition $UJ_pV = IT$. Now these properties are unaltered if the range space is enlarged. Therefore, we have:

Corollary 6.20. *Let $1 \leq p < \infty$. Every p-summing operator is both weakly compact and completely continuous.*

This has some striking implications for infinite dimensional spaces:

Corollary 6.21. *Let $1 \leq p < \infty$ and let X be an infinite dimensional Banach space. Then the identity operator on X is not p-summing.*

Proof. The composition of a completely continuous operator with a weakly compact operator is compact. □

In particular, taking $p = 1$, we see that every infinite dimensional Banach space X contains a sequence (x_n) such that the series $\sum_n \varphi(x_n)$ is absolutely convergent for every $\varphi \in X^*$, but $\sum_n \|x_n\|$ diverges. This is in marked contrast with the situation in finite dimensions, where unconditional and absolute convergence for series are the same. We have thus proved a weak form of the Dvoretzky–Rogers Theorem, which asserts that in every infinite dimensional Banach space there is an unconditionally convergent series that is not absolutely convergent.

When the domain of a p-summing operator is a $C(K)$ space, there is a particularly simple factorization:

Proposition 6.22. *Let $1 \le p < \infty$. An operator $T\colon C(K) \to Y$ is p-summing if and only if there exists a regular Borel probability measure μ on K and an operator $U\colon L_p(\mu) \to Y$ such that $T = UJ_p$. Furthermore, the p-summing norm of T is the minimum value of $\|U\|$ over all such factorizations:*

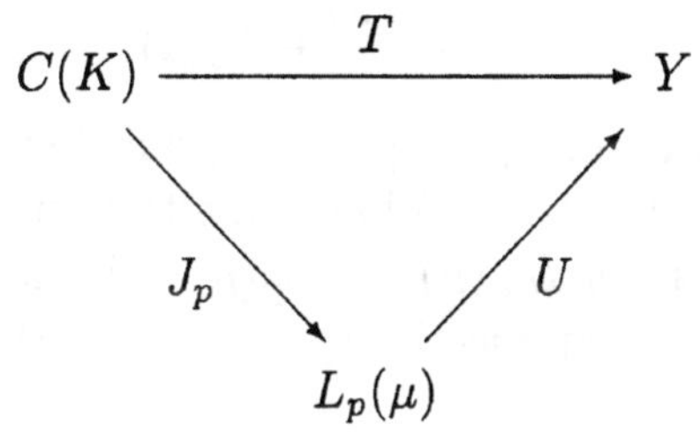

Proof. Since K is a weak*-compact norming subset of $C(K)^*$, it follows from the Pietsch Domination Theorem that there is a regular Borel probability measure μ on K, such that $\|Tx\|^p \le \pi_p(T) \int_K |x|^p \, d\mu$ for every $x \in C(K)$. Thus T is bounded for the $L_p(\mu)$ norm on $C(K)$ and it follows by the density of the continuous functions in $L_p(\mu)$ that there is an operator $U\colon L_p(\mu) \to Y$ such that $\|U\| \le \pi_p(T)$. On the other hand, $T = UJ_p$ implies that $\pi_p(T) \le \|U\|$ and so these norms are equal.

Conversely, if $T = UJ_p$ is any factorization of T, then $\pi_p(T) \le \|U\|$. □

The 2-summing operators have some special features that are unique to this case. We begin by looking again at the argument that produced the factorization scheme for p-summing operators. Recall the central idea: if $T\colon X \to Y$ is p-summing and K is a weak*-compact norming subset of B_{X^*}, then the Pietsch Domination Theorem provides a regular Borel probability measure, μ, such that $\|Tx\| \le \pi_p(T)\|J_pVx\|_p$, where V is the canonical embedding of X into $C(K)$ and $J_p\colon C(K) \to L_p(\mu)$ is the canonical mapping. We then define a bounded operator U on the subspace $J_pV(X)$ of $L_p(\mu)$ by $U(J_pVx) = Tx$. Our problem then was to extend the domain of V to all of $L_p(\mu)$ and this forced us to embed Y in an injective space. But when $p = 2$, the space $L_2(\mu)$ is a Hilbert space and there is an easier alternative.

We may assume that the operator U is defined on the closure of $J_2V(X)$. This subspace has an orthogonal complement in $L_2(\mu)$ and so we may extend U trivially by composing with the orthogonal projection of $L_2(\mu)$ onto the closure of $J_2V(X)$. Thus, we have:

Proposition 6.23. *Let X and Y be Banach spaces. An operator $T\colon X \to Y$ is 2-summing if and only if there exists a compact space K and a regular Borel probability measure μ on K, together with operators $V\colon X \to C(K)$ and $U\colon L_2(\mu) \to Y$, such that $T = UJ_2V$. Furthermore, $\pi_2(T) = \inf \|U\|\|V\|$, where the infimum is taken over all such factorizations, and this infimum is attained:*

$$\begin{array}{ccc} X & \xrightarrow{\quad T \quad} & Y \\ \Big\downarrow V & & \Big\uparrow U \\ C(K) & \xrightarrow{\quad J_2 \quad} & L_2(\mu) \end{array}$$

As was the case with integral operators, our factorizations through the canonical operators $C(K) \to L_p(\mu)$ have an alternative formulation in terms of the canonical operators $L_\infty(\mu) \to L_p(\mu)$. The advantage of the latter approach is that $L_\infty(\mu)$ is an injective space. We use this idea in the following result.

Proposition 6.24. *Let X and Y be Banach spaces and suppose that X is a subspace of the Banach space Z. Then every 2-summing operator from X into Y has an extension to a 2-summing operator from Z into Y with the same 2-summing norm.*

Proof. Consider a factorization of T of the type described in the proposition. The canonical mapping of $C(K)$ into $L_2(\mu)$ factors through the canonical mapping $J_2'\colon L_\infty(\mu) \to L_2(\mu)$. This gives the following factorization of T:

$$\begin{array}{ccc} X & \xrightarrow{\quad T \quad} & Y \\ V' \Big\downarrow & & \Big\uparrow U \\ L_\infty(\mu) & \xrightarrow{\quad J_2' \quad} & L_2(\mu) \end{array}$$

and $\pi_p(T) = \|U\|\|V'\|$. Since the space $L_\infty(\mu)$ is injective, the operator $V'\colon X \to L_\infty(\mu)$ can be extended to an operator $S\colon Z \to L_\infty(\mu)$ with the same norm. The operator UJ_pS is a 2-summing extension of T with the same 2-summing norm. □

This extension property of 2-summing operators has an interesting consequence for the tensor norm d_2. A tensor norm α is said to be *left injective* if, whenever X is a subspace of Z, then $X \otimes_\alpha Y$ is a subspace of $Z \otimes_\alpha Y$ for every Banach space Y. This is easily seen to be equivalent to the condition that every bounded linear functional on $X \otimes_\alpha Y$ has an extension to a bounded linear functional on $Z \otimes_\alpha Y$ with the same norm. Since $(X \hat{\otimes}_{d_2} Y)^* = \mathcal{P}_2(X, Y^*)$, we have:

Corollary 6.25. *The tensor norm d_2 is left injective.*

We conclude with an examination of the class of 2-summing operators on Hilbert spaces. Let H_1 and H_2 be Hilbert spaces. An operator $T\colon H_1 \to H_2$ is called a *Hilbert–Schmidt operator* if there exists an orthonormal basis $(e_i)_{i\in I}$ for H_1 such that $\sum_{i\in I} \|Te_i\|^2$ is finite. It is not difficult to show that this is then satisfied for every orthonormal basis for H_1 and that the value of the sum $\sum_{i\in I} \|Te_i\|^2$ is independent of the basis. The *Hilbert–Schmidt norm* of T is defined by

$$\sigma_2(T) = \left(\sum_{i\in I} \|Te_i\|^2\right)^{1/2}.$$

Proposition 6.26. *Let H_1 and H_2 be Hilbert spaces. An operator $T\colon H_1 \to H_2$ is 2-summing if and only if it is a Hilbert–Schmidt operator. Furthermore, the 2-summing and Hilbert–Schmidt norms coincide.*

Proof. Suppose that T is 2-summing. Let $(e_i)_{i\in I}$ be an orthonormal basis for H_1 and let $\{i_1, \dots, i_n\}$ be any finite subset of the indexing set I. It is easy to see that $\|(e_{i_k})_k\|_2^w = 1$. Therefore

$$\sum_{k=1}^n \|Te_{i_k}\|^2 \le \pi_2(T)^2.$$

Since this holds for every finite subset of I, it follows that T is a Hilbert–Schmidt operator and that $\sigma_2(T) \le \pi_2(T)$.

Conversely, suppose that T is a Hilbert–Schmidt operator. Fix orthonormal bases $(e_i)_{i\in I}$ and $(f_j)_{j\in J}$ for H_1 and H_2 respectively. Let $x_1, \dots, x_n \in H_1$. For each k, we have $x_k = \sum_i \langle x_k, e_i\rangle e_i$ and so the expansion of Tx_k in the basis (f_j) is given by $Tx_k = \sum_j \sum_i \langle x_k, e_i\rangle\langle Te_i, f_j\rangle f_j$. Therefore

$$\begin{aligned}\sum_k \|Tx_k\|^2 &= \sum_k \sum_j \left|\sum_i \langle x_k, e_i\rangle\langle Te_i, f_j\rangle\right|^2 \le \sum_k \sum_j \sum_i |\langle x_k, e_i\rangle|^2 |\langle Te_i, f_j\rangle|^2 \\ &= \sum_k |\langle x_k, e_i\rangle|^2 \sum_i \sum_j |\langle Te_i, f_j\rangle|^2 \le (\|(x_k)\|_2^w)^2\, \sigma_2(T)^2.\end{aligned}$$

Hence T is 2-summing and $\pi_2(T) \le \sigma_2(T)$. □

6.4 Grothendieck's Inequality

In this section, we prove a fundamental result, known as Grothendieck's Inequality, which will enable us to uncover some of the deeper facts about p-summing operators. The finite dimensional structure of the spaces involved is crucial and this will provide a further justification of the inclusion of the property of finite generation in the definition of a tensor norm.

Theorem 6.27 (Grothendieck's Inequality). *There is a positive constant, K, with the following property: for any n, let (a_{ij}) be an $n \times n$ matrix and let $x_1, \ldots, x_n$ and $y_1, \ldots, y_n$ be in the closed unit ball of a Hilbert space H. Then*

$$\left|\sum_{i,j} a_{ij}\langle x_i, y_j\rangle\right| \le K \sup\left\{\left|\sum_{i,j} a_{ij}s_i t_j\right| : |s_i|, |t_j| \le 1\right\}.$$

Proof. We give the proof for the case of real scalars. The complex case is dealt with by decomposition into real and imaginary parts. We may assume that the vectors x_i and y_i lie in an n-dimensional Hilbert space. Since all Hilbert spaces of a given dimension are isomorphic, we are free to work in any n-dimensional space. We take H to be the subspace of $L_2[0,1]$ spanned by the first n Rademacher functions $r_1, \ldots, r_n$. Thus $x_i(t) = \sum_{j=1}^n x_{ij} r_j(t)$ and $y_i(t) = \sum_{j=1}^n y_{ij} r_j(t)$ for $t \in [0,1]$.

We begin by truncating the functions $x_i(t)$ and $y_i(t)$. For any function x, we let

$$\overline{x}(t) = \begin{cases} \min\{x(t), 1\} & \text{if} \quad x(t) \ge 0, \\ \max\{x(t), -1\} & \text{if} \quad x(t) < 0. \end{cases}$$

Writing $\|a\|$ for $\sup\left\{\left|\sum_{i,j} a_{ij}s_i t_j\right| : |s_i|, |t_j| \le 1\right\}$, we have

$$\left|\sum_{i,j} a_{ij}\langle \overline{x_i}, \overline{y_j}\rangle\right| = \left|\sum_{i,j} a_{ij} \int_0^1 \overline{x_i}(t)\overline{y_j}(t)\, dt\right| \le \int_0^1 \left|\sum_{i,j} a_{ij}\overline{x_i}(t)\overline{y_j}(t)\right| dt \le \|a\|.$$

We now estimate the L_2 norm of the functions of the form $x - \overline{x}$. In general, if $x(t) \ne \overline{x}(t)$, then $|x(t) - \overline{x}(t)| = |x(t)| - 1$. Using the fact that $(\lambda - 2)^2 \ge 0$, we have $\lambda - 1 \le \lambda^2/4$ for every real number λ and so

$$|x(t) - \overline{x}(t)| \le x(t)^2/4 \tag{6.13}$$

for all such t. On the other hand, if $x(t) = \overline{x}(t)$, then (6.13) is also true. Therefore (6.13) holds for every function x and every t. Now suppose that $x = \sum_{j=1}^n b_j r_j$ with $\|x\| \le 1$. Then, using the orthogonality properties of the Rademacher functions established in Section 2.5 (see page 34),

$$
\begin{aligned}
16\int_0^1 |x(t)-\overline{x}(t)|^2\,dt &\le \int_0^1 x(t)^4\,dt \\
&= \int_0^1 \sum_{j,k,l,m=1}^{n} b_j b_k b_l b_m r_j(t) r_k(t) r_l(t) r_m(t)\,dt \\
&= \sum_{\substack{j=k,\\ l=m}} b_j b_k b_l b_m + \sum_{\substack{j=l,\\ k=m}} b_j b_k b_l b_m + \sum_{\substack{j=m,\\ k=l}} b_j b_k b_l b_m - 2\sum_j b_j^4 \\
&\le 3\Big(\sum_j b_j^2\Big)^2 \le 3\,.
\end{aligned}
$$

Therefore $\|x_i - \overline{x_i}\|$, $\|y_i - \overline{y_i}\| \le \sqrt{3}/4$ for each i. Let us write

$$|||a||| = \sup\left\{\left|\sum_{i,j} a_{ij}\langle x_i, y_j\rangle\right| : \|x_i\|, \|y_j\| \le 1\right\}.$$

Then

$$
\begin{aligned}
\left|\sum_{i,j} a_{ij}\langle x_i, y_j\rangle\right| &\le \left|\sum_{i,j} a_{ij}\langle \overline{x_i}, \overline{y_j}\rangle\right| + \left|\sum_{i,j} a_{ij}\langle x_i - \overline{x_i}, \overline{y_j}\rangle\right| \\
&\qquad + \left|\sum_{i,j} a_{ij}\langle x_i, y_j - \overline{y_j}\rangle\right| \\
&\le \|a\| + |||a|||\sqrt{3}/4 + |||a|||\sqrt{3}/4\,,
\end{aligned}
$$

and it follows that

$$|||a||| \le \frac{2}{2-\sqrt{3}}\|a\|$$

and we are done. □

The smallest constant that satisfies Grothendieck's Inequality is known as the *Grothendieck constant.* We shall denote this constant by K_G, or by $K_G(\mathbb{R})$ or $K_G(\mathbb{C})$ if it is necessary to distinguish between the real and complex cases. The proof given here does not give a good estimate for this constant. It is known that $\pi/2 < K_G(\mathbb{R}) \le 1.782$ and $1.338 \le K_G(\mathbb{C}) \le 1.405$.

We now present some striking applications of Grothendieck's Inequality.

Theorem 6.28. *Let X be an $\mathcal{L}_1$-space and let H be a Hilbert space. Then the norms d_p for $1 \le p \le \infty$ are all equivalent on $X \otimes H$.*

Proof. Since $d_\infty \le d_p \le d_1 = \pi$ for every p, it will suffice to show that $\pi \le C d_\infty$ for some constant C. Furthermore, since these norms are finitely generated, it will be enough to prove the result for $X = \ell_1^n$, with a constant that does not depend on n. We shall show that $\pi \le K_G d_\infty$ on $\ell_1^n \otimes H$.

Let $u \in \ell_1^n \otimes H$ with $d_\infty(u) < 1$. Then we may choose a representation $\sum_{i=1}^m x_i \otimes y_i$ of u such that $\|(x_i)\|_1^w < 1$ and $\|(y_i)\|_\infty < 1$. Let $x_i = \sum_{j=1}^n a_{ij} e_j$ for each j, where (e_j) is the standard unit vector basis for ℓ_1^n. We have

$$\begin{aligned}\|(x_i)\|_1^w &= \sup_{|s_i|\le 1} \Big\| \sum_{i=1}^n s_i x_i \Big\| = \sup_{|s_i|\le 1} \sup_{|t_j|\le 1} \Big| \sum_{j=1}^n t_j \Big(\sum_{i=1}^n s_i a_{ij} \Big) e_j \Big| \\ &= \sup \Big\{ \Big| \sum_{i,j} a_{ij} s_i t_j \Big| : |s_i|, |t_j| \le 1 \Big\} < 1\end{aligned}$$

To compute $\pi(u)$, let $T \in (\ell_1^n \hat{\otimes}_\pi H)^* = \mathcal{L}(\ell_1^n, H^*)$ with $\|T\| \le 1$. For each j, let $Te_j = z_j \in H$, so that $\|z_j\| \le 1$ and $\langle x, Te_j \rangle$ is given by the inner product $\langle x, z_j \rangle$. Then

$$\langle u, T \rangle = \sum_{i=1}^n \langle y_i, Tx_i \rangle = \sum_{i=1}^n \sum_{j=1}^n a_{ij} \langle y_i, Te_j \rangle = \sum_{i=1}^n \sum_{j=1}^n a_{ij} \langle y_i, z_j \rangle .$$

It follows from Grothendieck's Inequality that $|\langle u, T \rangle| \le K_G$ and hence $\pi(u) \le K_G$. Therefore $\pi(u) \le K_G d_\infty(u)$ for every $u \in \ell_1^n \otimes H$. □

Taking duals, we have:

Corollary 6.29. *Let X be an $\mathcal{L}_1$-space and let H be a Hilbert space. Then every operator $T \colon X \to H$ is 1-summing.*

In particular, we see that

$$\mathcal{L}(\ell_1, \ell_2) = \mathcal{P}_1(\ell_1, \ell_2) .$$

Our second application will require a little preparation.

Lemma 6.30. *Let (a_{ij}) be an $n \times n$ matrix and let $x_1, \ldots, x_n \in \ell_2^n$ satisfy $\|(x_i)\|_2^w \le 1$. Then*

$$\sum_k \Big(\sum_i \Big| \sum_j x_{ij} a_{jk} \Big|^2 \Big)^{1/2} \le K_G \sup \Big\{ \Big| \sum_{i,j} a_{ij} s_i t_j \Big| : |s_i|, |t_j| \le 1 \Big\} .$$

Proof. It follows from $\|(x_i)\|_2^w \le 1$ that $\sum_i |x_{ij}|^2 \le 1$ for each j. Thus the vectors $u_j = (x_{ij})_i$ lie in the closed unit ball of ℓ_2^n. For each k, we may choose a unit vector v_k in ℓ_2^n so that

$$\Big\| \sum_j a_{jk} u_j \Big\| = \Big\langle \sum_j a_{jk} u_j, v_k \Big\rangle .$$

Therefore

$$\sum_k\Big(\sum_i\Big|\sum_j x_{ij}a_{jk}\Big|^2\Big)^{1/2}=\sum_k\Big\|\sum_j a_{jk}u_j\Big\|=\sum_k\Big\langle\sum_j a_{jk}u_j,v_k\Big\rangle$$
$$=\sum_k\sum_j a_{jk}\langle u_j,v_k\rangle\le K_G\sup\Big\{\Big|\sum_{i,j}a_{ij}s_it_j\Big|:|s_i|,|t_j|\le 1\Big\}$$

by Grothendieck's Inequality. □

The class of $\mathcal{L}_p$-spaces for $1 < p < \infty$ is defined in exactly the same way as the $\mathcal{L}_1$ and $\mathcal{L}_\infty$-spaces. Thus, a Banach space X is an $\mathcal{L}_{p,\lambda}$-space for some $\lambda > 1$ if every finite dimensional subspace of X lies in a finite dimensional subspace that is isomorphic, via an operator T, with some ℓ_p^n, where $\|T\|\|T^{-1}\| \le \lambda$. A Banach space is said to be an *$\mathcal{L}_p$-space* if it is an $\mathcal{L}_{p,\lambda}$-space for some λ.

Theorem 6.31. *Let X be an $\mathcal{L}_\infty$-space and Y an $\mathcal{L}_p$-space where $1 \le p \le 2$. Then every operator $T\colon X \to Y$ is 2-summing.*

Proof. Thanks to the finitely dimensional characterization of the class of 2-summing operators given by Proposition 6.12 (a), it is enough to prove that if T is an operator from $X = \ell_\infty^m$ into $Y = \ell_p^m$, then $\pi_2(T) \le C\|T\|$, where the constant C is independent of the dimension m. We shall prove that $\pi_2(T) \le K_G\,\|T\|$.

We begin with the matrix representation of T relative to the standard bases (e_j) and (f_k) for ℓ_∞^m and ℓ_p^r respectively. Let $Te_j = \sum_k \alpha_{jk}f_k$ for each j. Then

$$Tx=\sum_k\sum_j x_j\alpha_{jk}f_k$$

for every $x \in \ell_\infty^m$ and so

$$\|Tx\|=\Big(\sum_k\Big|\sum_j x_j\alpha_{jk}\Big|^p\Big)^{1/p}.$$

Let $x_1,\dots,x_n \in \ell_\infty^m$ satisfy $\|(x_i)\|_2^w < 1$. We claim that

$$\Big(\sum_i\|Tx_i\|^2\Big)^{1/2}\le K_G\,\|T\|,$$

which is equivalent to

$$\Big(\sum_i\Big(\sum_k\Big|\sum_j x_{ij}\alpha_{jk}\Big|^p\Big)^{2/p}\Big)^{1/2}\le K_G\,\|T\|.$$

We now use the Lemma to establish a lower bound for $\|T\|$. If we view T as a bilinear form on $\ell_\infty^m \times \ell_{p'}^m$, we see that

$$\|T\| = \sup\left\{\left|\sum_{j,k} \alpha_{jk} s_j z_k\right| : \|s\|_\infty, \|z\|_{p'} \le 1\right\}.$$

Now fix a vector z in the closed unit ball of $\ell_{p'}^m$. Then $\|(t_k z_k)\|_{p'} \le 1$ for every t in the closed unit ball of ℓ_∞^m and it follows that

$$\sup\left\{\left|\sum_{j,k} \alpha_{jk} z_k s_j t_k\right| : |s_j|, |t_k| \le 1\right\} \le \|T\|.$$

Applying the Lemma to the matrix $(\alpha_{jk} z_k)$, we obtain

$$\begin{aligned}\sum_k \left(\sum_i \left|\sum_j x_{ij}\alpha_{jk} z_k\right|^2\right)^{1/2} &\le K_G \sup\left\{\left|\sum_{i,j} \alpha_{ij} z_j s_i t_j\right| : |s_i|, |t_j| \le 1\right\} \\ &\le K_G \|T\|.\end{aligned}$$

In other words,

$$\sum_k |z_k| \left(\sum_i \left|\sum_j x_{ij}\alpha_{jk}\right|^2\right)^{1/2} \le K_G \|T\|$$

for every z in the closed unit ball of $\ell_{p'}^m$. Taking the supremum over z, we have

$$\left(\sum_k \left(\sum_i \left|\sum_j x_{ij}\alpha_{jk}\right|^2\right)^{p/2}\right)^{1/p} \le K_G \|T\|.$$

If we write this as

$$\sum_k \left(\sum_i \left|\sum_j x_{ij}\alpha_{jk}\right|^2\right)^{p/2} \le (K_G \|T\|)^p$$

and observe that $2/p \ge 1$, then an application of Minkowski's Inequality yields

$$\left(\sum_i \left(\sum_k \left|\sum_j x_{ij}\alpha_{jk}\right|^p\right)^{2/p}\right)^{p/2} \le \sum_k \left(\sum_i \left|\sum_j x_{ij}\alpha_{jk}\right|^2\right)^{p/2} \le (K_G \|T\|)^p.$$

Therefore

$$\left(\sum_i \left(\sum_k \left|\sum_j x_{ij}\alpha_{jk}\right|^p\right)^{2/p}\right)^{1/2} \le K_G \|T\|,$$

and the proof is complete. □

Corollary 6.32. *Let X be an $\mathcal{L}_\infty$-space and Y an $\mathcal{L}_q$-space where $2 \le q \le \infty$. Then the norms d_p for $1 \le p \le 2$ are all equivalent on $X \otimes Y$.*

Proof. It follows from the proof of the proposition that $\pi \le K_G\, d_2$ on $\ell_\infty^m \otimes \ell_q^m$ for every m. It follows from the finitely generated nature of these tensor norms that, if X is an $\mathcal{L}_{\infty,\lambda}$-space and Y an $\mathcal{L}_{q,\mu}$-space, then $\pi \le \lambda\mu K_G\, d_2$ on $X \otimes Y$. □

There are some instances of this result worthy of note: on the tensor products $\ell_\infty \otimes \ell_\infty$ and $C(K) \otimes C(K)$, we have $\pi \le K_G\, d_2$. Indeed, we also have $\pi \le K_G\, g_2$. In fact, a sharper statement can be made. We shall return to this question in the next chapter.

6.5 Exercises

Exercise 6.1. Let X and Y be Banach spaces and $1 < p < \infty$. Does the formula $\|u\| = \inf\{\|(x_j)\|_p \|(y_j)\|_{p'} : u = \sum_{j=1}^n x_j \otimes y_j\}$ define a reasonable crossnorm on $X \otimes Y$?

Exercise 6.2. (a) Let $B \in \mathcal{B}(X \times Y)$ and let (x_α), (y_β) be bounded nets that converge in the weak* topology to $x^{**} \in X^{**}$, $y^{**} \in Y^{**}$ respectively. Show that the canonical left and right extensions of B are given by

$$ {}^*B(x^{**}, y^{**}) = \lim_\beta \lim_\alpha B(x_\alpha, y_\beta) $$

and

$$ B^*(x^{**}, y^{**}) = \lim_\alpha \lim_\beta B(x_\alpha, y_\beta) \,. $$

(b) Show that ${}^*B = B^*$ on $X^{**} \times Y$ and on $X \times Y^{**}$.

(c) Show that ${}^*B = B^*$ on $X^{**} \times Y^{**}$ if and only if the operator $T \colon X \to Y^*$ associated with B is weakly compact.

Exercise 6.3. Show that $d_p = g_p$ on $\ell_p \otimes \ell_p$.

Exercise 6.4. Let (Ω, Σ, μ) be a finite measure space. For $1 < p < \infty$, let $L_p(\mu, X)$ be the space of equivalence classes of μ-measurable functions $f \colon \Omega \to X$ for which the function $\|f(\cdot)\|^p$ is integrable, with the norm $\|f\|_p = (\int_\Omega \|f\|^p \, d\mu)^{1/p}$. Show that

$$ d_p(f) \le \|f\|_p \le g_p(f) $$

for every f belonging to the subspace $L_p(\mu) \otimes X$ of $L_p(\mu, X)$.

Exercise 6.5. Let $1 \le p < \infty$ and let $\lambda \in \ell_\infty$. Show that the diagonal operator $D_\lambda \colon c_0 \to c_0$ given by $D_\lambda x = (\lambda_n x_n)$ is p-summing if and only if $\lambda \in \ell_p$.

Exercise 6.6. Let μ be a positive regular Borel measure on the compact space K and let $g \in L_p(\mu)$. The multiplication operator $M_g \colon C(K) \to L_p(\mu)$ is defined by $M_g f = gf$. Show that M_g is p-summing and calculate $\pi_p(M_g)$.

Exercise 6.7. Show that every integral operator T is 1-summing and that $\pi_1(T) \leq \|T\|_I$. Find an example of a 1-summing operator that is not integral.

Exercise 6.8. Show that $\mathcal{P}_1(c_0, Y) = \mathcal{N}(c_0, Y)$ for every Banach space Y, with equality of the 1-summing and nuclear norms.

Exercise 6.9. Let (Ω, Σ, μ) be a finite measure space and let $T\colon X \to Y$ be a 1-summing operator. If $f\colon \Omega \to X$ is a μ-measurable Pettis integrable function, show that the function $Tf\colon \Omega \to Y$ is Bochner integrable.

Exercise 6.10. Show that the space $\mathcal{P}_p(X, Y)$ is complete in the p-summing norm.

Exercise 6.11. Show that the closed subspace of $L_1(\mu)$ spanned by the Rademacher functions is not complemented.

Exercise 6.12. Show that every $L_p(\mu)$ is an $\mathcal{L}_p$-space.

Exercise 6.13. Show that every $\mathcal{L}_2$-space is isomorphic to a Hilbert space (Hint: there is a quotient operator from some $\ell_1(I)$ onto the $\mathcal{L}_2$-space X. This operator is 2-summing; now consider the factorization given by Proposition 6.23.)

Exercise 6.14. Let X be an $\mathcal{L}_p$-space, where $1 \leq p \leq 2$. Show that every weakly summable sequence in X is strongly 2-summable (Hint: if (x_n) is a weakly summable sequence in X, consider the operator $T\colon c_0 \to X$ given by $Te_n = x_n$.)

7. Tensor Norms

In this chapter, we study the basic properties of tensor norms. We begin with the dual norm and this leads naturally to the vital concept of accessibility, which can be thought of as an analogue for tensor norms of the approximation property for spaces. We then meet the various injective and projective norms that can be associated with a tensor norm. Next, we turn our attention to the identification of the duals of the Chevet–Saphar tensor norms in terms of the classes of p-integral operators. In the final section, we meet the Hilbertian tensor norm, which plays a central role in the theory. Grothendieck's Inequality can now be interpreted as the statement that the Hilbertian tensor norm and the largest injective norm are equivalent. Two new classes of operators emerge: the Hilbertian and the 2-dominated operators. We conclude with Grothendieck's classification of the natural tensor norms.

7.1 The Dual Norm

In finite dimensions, the definition of the dual norm is clear. If E and F are finite dimensional normed spaces and α is a tensor norm, then $E \otimes F$ is, algebraically, the dual space of $E^* \otimes_\alpha F^*$ and we may define α' to be the dual norm:

$$E \otimes_{\alpha'} F = (E^* \otimes_\alpha F^*)^* .$$

In other words, if $u \in E \otimes F$, then

$$\alpha'(u) = \sup\{|\langle u, v\rangle| : v \in E^* \otimes F^*,\ \alpha(v) \leq 1\} .$$

To spell this out, recall that the duality between $E \otimes F$ and $E^* \otimes F^*$ works in the following way: if $u = \sum_{i=1}^n x_i \otimes y_i \in E \otimes F$ and $v = \sum_{j=1}^m \varphi_j \otimes \psi_j \in E^* \otimes F^*$, then $\langle u, v\rangle = \sum_{i,j} \varphi_j(x_i)\psi_j(y_i)$. Alternatively, we can use trace duality: if we view u and v as operators from E^* into F and F into E^* respectively, then $\langle u, v\rangle = \operatorname{tr} vu$.

It is tempting to use the same approach in infinite dimensions. If X and Y are Banach spaces, consider the dual space $(X^* \otimes_\alpha Y^*)^*$. While it is no longer the case that this dual coincides with $X \otimes Y$, there is a canonical algebraic embedding of $X \otimes Y$ into it. Thus, the dual norm of $(X^* \otimes_\alpha Y^*)^*$ induces a norm on $X \otimes Y$. We shall refer to this norm as the *Schatten dual norm* and we denote it by α^s. To summarize, we have an embedding

$$X \otimes_{\alpha^s} Y \subset (X^* \otimes_\alpha Y^*)^*$$

and, just as in finite dimensions, if $u \in X \otimes Y$, then

$$\alpha^s(u) = \sup\{|\langle u, v\rangle| : v \in X^* \otimes Y^*,\ \alpha(v) \leq 1\}.$$

Let us apply this definition to the projective and injective norms. $X \otimes_\varepsilon Y$ is a subspace of $(X^* \otimes_\pi Y^*)^*$ and so $\pi^s = \varepsilon$. But we have seen in Chapter 4 that the norm induced on $X \otimes Y$ by $(X^* \otimes_\varepsilon Y^*)^*$ is not, in general, the projective norm. Indeed, by Proposition 4.14, the canonical embedding of $X \otimes_\pi Y$ into $(X^* \otimes_\varepsilon Y^*)^*$ is an isometric embedding for every Banach space Y if and only if X has the metric approximation property. The existence of a space without the metric approximation property shows that there is at least one tensor product on which $\varepsilon^s < \pi$, or, to put it another way, $(\pi^s)^s < \pi$. This is certainly not what we expect.

The flaw in the Schatten dual is not difficult to detect. While α^s is easily seen to be a uniform crossnorm, it fails to possess the crucial property of finite generation. The remedy is simple; we use the extension process given in Proposition 6.3 to extend the definition of the dual from finite dimensions. Thus, we define the *dual norm*, α', to be the unique tensor norm that agrees with the dual norm on tensor products of finite dimensional spaces. So, for $u \in X \otimes Y$, we have

$$\alpha'(u) = \inf\{\alpha'_{E,F}(u) : u \in E \otimes F,\ \dim E, \dim F < \infty\}, \tag{7.1}$$

the infimum being taken over all pairs of finite dimensional subspaces of X and Y whose tensor product contains u. With this definition, all the properties we would hope for follow easily from the finite dimensional case. So, for example, we have $\pi' = \varepsilon$ and $\varepsilon' = \pi$. And, of course, $(\alpha')' = \alpha$ for every tensor norm α.

The relationship between the dual norm and the Schatten dual is of some interest. While the definition of α' is the "correct" one, the norm α^s is easier to understand and to compute. We have $\alpha^s \leq \alpha'$ in general and these norms coincide in finite dimensions. It follows that $\alpha^s = \alpha'$ on $X \otimes Y$ if and only if α^s is finitely generated on $X \otimes Y$. We have shown in Proposition 6.2 that every uniform crossnorm is finitely generated on the tensor product of two spaces with the metric approximation property. Thus we have

Proposition 7.1. *Let X and Y be Banach spaces with the metric approximation property. Then $\alpha^s = \alpha'$ on $X \otimes Y$.*

This result does not explain the fact that $\pi^s = \pi' = \varepsilon$. This coincidence can be explained by the possession by the injective norm of a property that is dual to finite generation.

Let E and F be closed subspaces of the Banach spaces X and Y and let $Q_E\colon X \to X/E$ and $Q_F\colon Y \to Y/F$ be the quotient operators. If α is a uniform crossnorm then

$$\alpha((Q_E \otimes Q_F)u; (X/E) \otimes (Y/F)) \le \alpha(u)$$

for every $u \in X \otimes Y$. A uniform crossnorm α is said to be *cofinitely generated* if, for every pair of Banach spaces X, Y and every $u \in X \otimes Y$, we have

$$\alpha(u) = \sup\{\alpha((Q_E \otimes Q_F)u; (X/E) \otimes (Y/F)) : \dim X/E, \dim Y/F < \infty\}.$$

In other words, the α-norm of a tensor u is determined by projecting u onto finite dimensional tensor products and computing the norm there. It is not difficult to show that the injective tensor norm ε is cofinitely generated, although this will also follow from our next result. It may come as a surprise to find that the projective norm is not, in spite of its projective behaviour.

Proposition 7.2. *Let α be a tensor norm.*

(a) *The Schatten dual norm α^s is a cofinitely generated uniform crossnorm.*
(b) *$\alpha^s = \alpha'$ if and only if the dual norm α' is cofinitely generated.*

Proof. (a) Let $u \in X \otimes Y$ and let $\varepsilon > 0$. By the definition of the norm α^s, there exists $v \in X^* \otimes Y^*$ such that $\alpha(v) < 1$ and $\langle u, v\rangle > \alpha^s(u) - \varepsilon$. Since α is a tensor norm, there exist finite dimensional subspaces M and N of X^* and Y^* respectively such that $v \in M \otimes N$ and $\alpha(v; M \otimes N) < 1$. There exist subspaces E and F of X and Y respectively such that $M = (X/E)^*$ and $N = (Y/F)^*$. Then

$$\alpha^s((Q_E \otimes Q_F)u; (X/E) \otimes (Y/F)) \ge \langle u, v\rangle \ge \alpha^s(u) - \varepsilon.$$

It follows that α^s is cofinitely generated.

(b) If $\alpha^s = \alpha'$ then it follows from (a) that α' is cofinitely generated. Conversely, suppose that α' is cofinitely generated. Then, since α^s and α' coincide on tensor products of finite dimensional spaces, we have $\alpha^s = \alpha'$.

□

Corollary 7.3. *The projective norm π is not cofinitely generated.*

We obtain another viewpoint on these phenomena if we interchange the roles of α and α'. The canonical embedding of $X \otimes Y$ into $(X^* \otimes_{\alpha'} Y^*)^*$ induces the Schatten dual norm $(\alpha')^s$ on $X \otimes Y$. On the other hand, we have the norm $\alpha'' = \alpha$ on this space. Then $(\alpha')^s \le \alpha$ and, in general, these norms are distinct. To keep our notation managable, we shall write

$$\tilde{\alpha} = (\alpha')^s$$

and we shall refer to $\tilde{\alpha}$ as the *associated norm* of α. Thus, we have an isometric embedding:

$$X \otimes_{\tilde{\alpha}} Y \subset (X^* \otimes_{\alpha'} Y^*)^* . \tag{7.2}$$

When the spaces in question are duals, the norm $\tilde{\alpha}$ can also be defined by a simpler embedding. It follows from Theorem 6.5 that the same norm is induced on $X^* \otimes Y^*$ by $(X^{**} \otimes_{\alpha'} Y^{**})^*$ and $X \otimes_{\alpha'} Y$. Thus we have another isometric embedding:

$$X^* \otimes_{\tilde{\alpha}} Y^* \subset (X \otimes_{\alpha'} Y)^* . \tag{7.3}$$

The tensor norm α is said to be *totally accessible* if $\tilde{\alpha} = \alpha$. By Proposition 7.2, this is equivalent to α being cofinitely generated. Our previous results can now be phrased as follows: the injective norm ε is totally accessible, and the projective norm π is not.

Though the projective norm is not totally accessible, all is not lost. By Theorem 4.14, we do have an isometric embedding of $X \otimes_\pi Y$ into $(X^* \otimes_\varepsilon Y^*)^*$ if just *one* of the spaces X, Y has the metric approximation property. We shall see that this property is shared by a great many tensor norms and so we give it a name. The tensor norm α is said to be *accessible* if the norms $\tilde{\alpha}$ and α are equal on $X \otimes Y$ whenever either X or Y has the metric approximation property. We leave it as an exercise to the reader that this is equivalent to the condition that $\tilde{\alpha} = \alpha$ on $X \otimes Y$ whenever either X or Y is finite dimensional. Accessibility can be thought of an approximation property for tensor norms. Indeed, Grothendieck, who coined this terminology, applied the term "accessible" both to spaces and to tensor norms. All the tensor norms we encounter in this book are accessible. The existence of a non-accessible tensor norm was established in 1991 by Pisier.

We can summarize our findings as follows. In order that the norm induced on $X \otimes Y$ by the embedding into $(X^* \otimes_{\alpha'} Y^*)^*$ coincides with α, one of the following conditions suffices:

(a) Both X and Y have the metric approximation property.
(b) α is totally accessible.
(c) α is accessible and either X or Y has the metric approximation property.

We collect some useful results concerning accessibility:

Proposition 7.4. (a) *Every injective tensor norm is totally accessible.*
(b) *Every projective tensor norm is accessible.*
(c) *A tensor norm α is accessible if and only if the dual norm α' is accessible.*
(d) *If the tensor norm α is totally accessible then α' is accessible.*

Proof. (a) Since $\tilde{\alpha} \le \alpha$ holds in general, we must prove the reverse inequality. Let X and Y be Banach spaces and let $J_X : X \to C(B_{X^*})$ and $J_Y : Y \to C(B_{Y^*})$ be the canonical embeddings. By the uniform property of $\tilde{\alpha}$, the tensor product operator

$$J_X \otimes J_Y : X \otimes_{\tilde{\alpha}} Y \to C(B_{X^*}) \otimes_{\tilde{\alpha}} C(B_{Y^*})$$

has norm at most one. Let $u \in X \otimes Y$. Then

$$\tilde{\alpha}((J_X \otimes J_Y)u; C(B_{X^*}) \otimes C(B_{Y^*})) \leq \tilde{\alpha}(u; X \otimes Y).$$

Now every $C(K)$ space has the metric approximation property and so $\tilde{\alpha} = \alpha$ on $C(B_{X^*}) \otimes C(B_{Y^*})$. Therefore

$$\alpha((J_X \otimes J_Y)u; C(B_{X^*}) \otimes C(B_{Y^*})) \leq \tilde{\alpha}(u; X \otimes Y).$$

But, by the injectivity of α, the α-norm of u in $X \otimes Y$ is the same as the α-norm of $(J_X \otimes J_Y)u$. Hence $\alpha(u; X \otimes Y) \leq \tilde{\alpha}(u; X \otimes Y)$ and we are done.

(b) If α is projective then α' is injective (see Proposition 7.5 below) and the result follows from (a) and (d).

(c) To show that α' is accessible, it will suffice to prove that $E \otimes_{\alpha'} Y$ is a subspace of $(E^* \otimes_\alpha Y^*)^*$ for every finite dimensional space E and every Y. By the finite dimensionality of E, the dual space of $E \otimes_{\alpha'} Y$ coincides algebraically with $E^* \otimes Y^*$. Hence, by the accessibility of α, we have $(E \otimes_{\alpha'} Y)^* = E^* \otimes_\alpha Y^*$. Therefore $E \otimes_{\alpha'} Y \subset (E \otimes_{\alpha'} Y)^{**} = (E^* \otimes_\alpha Y^*)^*$ and the proof is complete. (d) follows from (c). □

We note that the dual of a totally accessible norm need not be totally accessible. The injective norm provides an example.

We have seen the importance of injectivity and projectivity. Our next result shows that, as one would expect, these are dual properties.

Proposition 7.5. *A tensor norm α is projective if and only if the dual norm α' is injective.*

Proof. Suppose that α is projective. Let X and Y be closed subspaces of W and Z respectively. Let $u \in X \otimes Y$ and let $\varepsilon > 0$. If E and F are finite dimensional subspaces of X and Y respectively such that $u \in E \otimes F$, then $\alpha'(u; X \otimes Y) \leq \alpha'(u; E \otimes F)$. The dual space of $E \otimes_{\alpha'} F$ is $E^* \otimes_\alpha F^*$ and hence there exists $v \in E^* \otimes F^*$ such that $\alpha(v; E^* \otimes F^*) < 1$ and

$$\alpha'(u; E \otimes F) \leq \langle u, v \rangle + \varepsilon.$$

Let $P\colon W^* \to E^*$ and $Q\colon Z^* \to F^*$ be the canonical quotient operators. Then $P \otimes Q\colon W^* \otimes_\alpha Y^* \to E^* \otimes_\alpha F^*$ is a quotient operator and so there exists $w \in W^* \otimes_\alpha Z^*$ such that $\alpha(w; W^* \otimes Z^*) < 1$ and $(P \otimes Q)w = v$. Then $\langle u, v \rangle = \langle u, w \rangle$ and so, by the definition of α^s, we have

$$\begin{aligned}\alpha'(u; X \otimes Y) &\leq \alpha'(u; E \otimes F) \leq \langle u, v \rangle + \varepsilon = \langle u, w \rangle + \varepsilon \\ &\leq \alpha^s(u; W \otimes Z) + \varepsilon \leq \alpha'(u; W \otimes Z) + \varepsilon\end{aligned}$$

Therefore $\alpha'(u; X \otimes Y) \leq \alpha'(u; W \otimes Z)$ and it follows that α' is injective.

Conversely, suppose that α' is injective. Let $P\colon W \to X$ and $Q\colon Z \to Y$ be quotient operators. We treat first the case in which all the spaces in question are finite dimensional. By the injectivity of α', the space $X^* \otimes_{\alpha'} Y^*$ is a subspace of $W^* \otimes_{\alpha'} Z^*$. Hence $X \otimes_\alpha Y = (X^* \otimes_{\alpha'} Y^*)^*$ is a quotient space of $W \otimes_\alpha Z = (W^* \otimes_{\alpha'} Z^*)^*$.

We now consider the general case. Let $u \in X \otimes Y$ and let $\varepsilon > 0$. It will suffice to show that if $\alpha'(u; X \otimes Y) = 1$, then there exists $v \in W \otimes Z$ such that $(P \otimes Q)v = u$ and $\alpha'(v; W \otimes Z) \leq (1+\varepsilon)^4$. We begin by choosing finite dimensional subspaces E and F of X and Y respectively such that

$$\alpha'(u; E \otimes F) \leq (1+\varepsilon)\alpha'(u; X \otimes Y).$$

Using a compactness argument, there exists a finite dimensional subspace G of W such that P maps G onto E and, for every $x \in E$, there exists $w \in G$ such that $Pw = x$ and $\|x\| \leq \|w\| \leq (1+\varepsilon)\|x\|$. Let E_0 be the space E with the norm given by the Minkowski functional of the set $P(B_G)$. Then $P\colon G \to E_0$ is a quotient operator and the identity mapping from E to E_0 has norm at most $(1+\varepsilon)$. We carry out an analogous construction for the space F, obtaining a "close" renorming F_0 of F which is a quotient of some finite dimensional subspace H of Z. By the uniform property of α, we have

$$\alpha'(u; E_0 \otimes F_0) \leq (1+\varepsilon)^2\alpha'(u; E \otimes F).$$

Now $E_0 \otimes_{\alpha'} F_0$ is a quotient of $G \otimes_{\alpha'} H$ and so there exists $v \in G \otimes H$ such that $(P \otimes Q)v = u$ and

$$\alpha'(v; G \otimes H) \leq (1+\varepsilon)\alpha'(u; E_0 \otimes F_0).$$

Combining these inequalities, we have

$$\alpha'(v; W \otimes Z) \leq \alpha'(v; G \otimes H) \leq (1+\varepsilon)^3\alpha'(u; E \otimes F) \leq (1+\varepsilon)^4\alpha'(u; X \otimes Y).$$

This concludes the proof. □

An inspection of this proof shows that, in order to establish the projectivity or injectivity of a tensor norm, it suffices to show that the relevant property holds for tensor products of finite dimensional spaces.

We now turn our attention to the order structure of the set of tensor norms. Not only is there a biggest and a smallest tensor norm; we see now that every non-empty set of tensor norms has a greatest lower bound and a least upper bound.

Proposition 7.6. *The set of tensor norms is a complete lattice.*

Proof. Let $\{\alpha_i : i \in I\}$ be a non-empty set of tensor norms. We define the least upper bound first in finite dimensions. If E and F are finite dimensional, let

$$\alpha(u; E \otimes F) = \sup_{i \in I} \alpha_i(u; E \otimes F)$$

for $u \in E \otimes F$. It is easy to see that α is a uniform crossnorm on the class of tensor products of finite dimensional spaces and that α is the smallest such norm that satisfies $\alpha_i \leq \alpha$ for every $i \in I$. The unique extension of α to a tensor norm as defined in Proposition 6.3 is the least upper bound of the set $\{\alpha_i : i \in I\}$.

To construct the greatest lower bound, consider the set $\{\alpha_i' : i \in I\}$. Let β be the least upper bound of this set, so that β is the smallest tensor norm that satisfies $\alpha_i' \leq \beta$ for every $i \in I$. Then β' is the biggest tensor norm that satisfies $\beta' \leq \alpha_i'$ for every $i \in I$ and hence β' is the greatest lower bound. □

A word of warning is in order here. It is tempting, but rather misleading, to write $\sup_i \alpha_i$ and $\inf_i \alpha_i$ for the least upper bound and greatest lower bound.

7.2 Injective and Projective Associates

We wish to determine the duals of the Chevet–Saphar tensor norms d_p and g_p. We have seen that the accessibility, or total accessibility of a tensor norm greatly simplifies the determination of the dual. Now d_p and g_p are neither injective nor projective for $1 < p \leq \infty$ and so the results of the previous section cannot be applied. However, d_p is right projective and, even better, d_2 is also left injective. We take this observation as our starting point.

We recall that a tensor norm α is right projective if, whenever $P\colon X \to Y$ is a quotient operator, then the tensor product operator

$$I \otimes P\colon Z \otimes_\alpha X \to Z \otimes_\alpha Y$$

is a quotient operator for every Banach space Z. There are analogous definitions for left projectivity and for left and right injectivity. By the remark following the proof of Proposition 7.5, in order to show that a tensor norm has any one of these properties, it suffices to establish the property for tensor products of finite dimensional spaces. Proceding as in the proof of Proposition 7.5, we see that α is right (respectively, left) injective if and only if the dual norm α' is right (respectively, left) projective.

The notion of accessibility can also be developed in a one-sided fashion. Recall that a tensor norm α is called accessible if the canonical embedding of $X \otimes Y$ into $(X^* \otimes_{\alpha'} Y^*)^*$ induces the norm α on $X \otimes Y$ provided X or Y has the metric approximation property. In other words, we have $(\alpha')^s = \alpha$ in the presence of the metric approximation property. Similarly, accessibility of α' guarantees $\alpha^s = \alpha'$ under the same conditions.

A tensor norm α is said to be *left accessible* if the canonical embedding of $X \otimes Y$ into $(X^* \otimes_{\alpha'} Y^*)^*$ induces the norm α on $X \otimes Y$ whenever the Banach space Y has the metric approximation property. Similarly, α is said to be *right accessible* if $X \otimes_\alpha Y$ is a subspace of $(X^* \otimes_{\alpha'} Y^*)^*$ whenever X has the metric approximation property. In each of these definitions, the assumption of the metric approximation property can be replaced by the assumption of finite dimensionality. Clearly, a tensor norm α is accessible if and only if it is both left and right accessible. The following proposition is proved in exactly the same way as Proposition 7.4.

Proposition 7.7. *Let α be a tensor norm.*

(a) *If α is left injective or left projective, then α is left accessible.*
(b) *α is left accessible if and only if α' is left acccessible.*

One may, of course, replace "left" by "right" in the statement of this result. It follows that d_p is right accessible and d_2 is accessible. We shall see that much more is true for these norms.

We now describe a process by means of which an arbitrary tensor norm can be modified so that the new norm has any desired property of left or right injectivity or projectivity. We begin with right injectivity.

Proposition 7.8. *For every tensor norm α there exists a biggest right injective tensor norm β such that $\beta \leq \alpha$.*

Proof. Let $\mathcal{R}$ be the set of tensor norms that are right injective and bounded above by α. Then $\mathcal{R}$ is non-empty, since it contains the injective norm. Let β be the least upper bound of this set. In order to show that β is right injective, it suffices to consider finite dimensional spaces. Suppose that X, Y and Z are finite dimensional and Y is a subspace of Z. Let $u \in X \otimes Y$. Then, by the definition of the least upper bound, we have

$$\beta_{X,Y}(u) = \sup_{\gamma \in \mathcal{R}} \gamma_{X,Y}(u) = \sup_{\gamma \in \mathcal{R}} \gamma_{X,Z}(u) = \beta_{X,Z}(u)\,.$$

Therefore β is right injective. □

The tensor norm β is known as the *right injective associate* of α. We shall use the notation $\alpha\backslash$ for this norm. The same construction can be carried out on the left to obtain the *left injective associate*, which we denote by $/\alpha$. Thus $/\alpha$ is the biggest left injective tensor norm that is bounded above by α. If α is already left or right injective, then the appropriate injective associate norm coincides with α.

These constructions can be combined, giving the norms $/(\alpha\backslash)$ and $(/\alpha)\backslash$. We leave it as an easy exercise to show that these norms are equal. Therefore, we can define

$$/\alpha\backslash = /(\alpha\backslash) = (/\alpha)\backslash\,.$$

The norm $/\alpha\backslash$ is the biggest injective norm that is bounded above by α. It is known as the *injective associate* of α. The injective associate of α is the biggest injective norm that is bounded above by α. In particular, $/\pi\backslash$ is the biggest injective tensor norm.

Our definition of the injective associates, while elegant, can be difficult to apply. The following proposition gives a more practical characterization.

Proposition 7.9. *Let α be a tensor norm.*

(a) *Let X and Y be Banach spaces. The norm induced on $X \otimes Y$ by the canonical embedding of $X \otimes Y$ into $X \otimes_\alpha C(B_{Y^*})$ is $\alpha\backslash$.*
(b) *For every Banach space X and every compact space K, the norms α and $\alpha\backslash$ coincide on $X \otimes C(K)$.*
(c) *If Z is an injective space then the norms α and $\alpha\backslash$ coincide on $X \otimes Z$ for every Banach space X.*

Proof. (a) For every pair X, Y of Banach spaces, let β be the norm on $X \otimes Y$ induced by the canonical embedding $X \otimes Y \to X \otimes_\alpha C(B_{Y^*})$. It is easy to see that β is a uniform crossnorm that satisfies $\beta \leq \alpha$, and and is finitely generated on the left, in the sense that

$$\beta_{X,Y}(u) = \inf\{\beta_{E,Y}(u) : u \in E \otimes Y,\ \dim E < \infty\}.$$

Since α is a tensor norm, $X \otimes_\alpha C(B_{Y^*})$ is a subspace of $X \otimes_\alpha C(B_{Y^*})^{**}$ by Proposition 6.4. Therefore, β is the norm induced by the canonical embedding $X \otimes Y \to X \otimes_\alpha C(B_{Y^*})^{**}$. It follows that β is right injective. Indeed, suppose that Y is a subspace of the Banach space Z. We have $X \otimes_\beta Y \subset X \otimes_\alpha C(B_{Y^*})^{**}$ and $X \otimes_\beta Z \subset X \otimes_\alpha C(B_{Z^*})^{**}$. But the injective space $C(B_{Y^*})^{**}$ is complemented in $C(B_{Z^*})^{**}$ by a projection of unit norm and so $X \otimes_\alpha C(B_{Y^*})^{**}$ is a subspace of $X \otimes_\alpha C(B_{Z^*})^{**}$. Therefore $X \otimes_\beta Y$ is a subspace of $X \otimes_\beta Z$. It follows from the right injectivity of β that this norm is finitely generated on the right. Hence β is a right injective tensor norm. Since β is bounded above by α, we have $\beta \leq \alpha\backslash$.

On the other hand, let γ be any right injective tensor norm that is bounded above by α. Then $\gamma \leq \alpha$ on $X \otimes C(B_{Y^*})$ and it follows from the definition of β that $\gamma \leq \alpha$ on every tensor product $X \otimes Y$. Therefore β is the biggest right injective tensor norm that is bounded above by α.

(b) Let $Y = C(K)$ for some compact space K. Then Y is complemented in the space $C(B_{Y^*})$ by the unit norm projection $f \mapsto \tilde{f}$, where $\tilde{f}(t) = f(\delta_t)$. It follows that $X \otimes_\alpha Y$ is a subspace of $X \otimes_\alpha C(B_{Y^*})$ and so, by the definition of $\beta = \alpha\backslash$ in the proof of (a), $\alpha\backslash = \alpha$ on $X \otimes Y$.

(c) follows from (a). □

This proposition provides a practical way to approach the injective associates. We have

$$X \otimes_{\alpha\backslash} Y \subset X \otimes_\alpha C(B_{Y^*}). \tag{7.4}$$

Alternatively, we may embed Y in any injective space; for example, we may take the canonical embedding of Y into the space $\ell_\infty(B_{Y^*})$, to get

$$X \otimes_{\alpha\backslash} Y \subset X \otimes_\alpha \ell_\infty(B_{Y^*}). \tag{7.5}$$

Similar descriptions apply to the left injective associate $/\alpha$ and the injective associate $/\alpha\backslash$.

We now turn our attention to the construction of the associated projective norms.

Proposition 7.10. *For every tensor norm α there exists a smallest right projective tensor norm β such that $\beta \geq \alpha$.*

Proof. Let $\beta = (\alpha'\backslash)'$. Since $\beta \geq \alpha$ if and only if $\beta' \leq \alpha'$ and a tensor norm is right projective if and only if its dual is right injective, it follows that β has the required properties. □

We shall denote this tensor norm by $\alpha/$ and we shall refer to it as the *right projective associate* of α. In the same way, we have the *left projective associate*, $\backslash\alpha$ and the *projective associate*, $\backslash\alpha/$, where

$$\backslash\alpha/ = \backslash(\alpha/) = (\backslash\alpha)/ .$$

In particular, the norm $\backslash\varepsilon/$ is the smallest projective tensor norm. Great care must be taken when combining injective and projective associates. For example, it is not true in general that $\backslash(\alpha\backslash)$ is the same as $(\backslash\alpha)\backslash$. Indeed, we shall see that the case $\alpha = g_p$ provides a counterexample.

There is a projective analogue of Proposition 7.9 that provides a useful insight into the uses of the projective associate norms. The kernel of this result lies, as always in these matters, in a finite dimensional statement:

Lemma 7.11. *Let α be a tensor norm. If E is finite dimensional then $\alpha/ = \alpha$ on $E \otimes \ell_1^n$ for every n.*

Proof. Using the definition of $\alpha/$, the fact that ℓ_∞^n is a projective space, and the definition of the dual norm in finite dimensions, we have the following chain of isometric isomorphisms:

$$E \otimes_{\alpha/} \ell_1^n = E \otimes_{(\alpha'\backslash)'} \ell_1^n = (E^* \otimes_{\alpha'\backslash} \ell_\infty^n)^* = (E^* \otimes_{\alpha'} \ell_\infty^n)^* = E \otimes_\alpha \ell_1^n .$$

□

It follows that

$$X \otimes_{\alpha/} \ell_1(I) = X \otimes_\alpha \ell_1(I)$$

for every Banach space X and every indexing set I. More than this is true: $\ell_1(I)$ can be replaced by any space that is a $\mathcal{L}_{1,\lambda}$-space for every $\lambda > 1$.

Proposition 7.12. *Let α be a tensor norm.*

(a) *Let X and Y be Banach spaces and let $P\colon \ell(I) \to Y$ be a quotient operator. Then the norm $\alpha/$ on $X \otimes Y$ coincides with the quotient norm defined by the operator $I \otimes P\colon X \otimes_\alpha \ell_1(I) \to X \otimes Y$.*

(b) *If Y is a $\mathcal{L}_{1,\lambda}$-space for every $\lambda > 1$, then the norms α and $\alpha/$ coincide on $X \otimes Y$ for every Banach space X.*

(c) *For every Banach space X and every finite, positive measure μ, the norms α and $\alpha/$ coincide on $X \otimes L_1(\mu)$.*

Proof. (a) Since $I \otimes P\colon X \otimes_{\alpha/} \ell_1(I) \to X \otimes_{\alpha/} Y$ is a quotient operator and $\ell_1(I)$ is a $\mathcal{L}_{1,\lambda}$-space for every $\lambda > 1$, the result follows from (b).

(b) follows from the Lemma.

(c) follows from (b), since $L_1(\mu)$ is a $\mathcal{L}_{1,\lambda}$-space for every $\lambda > 1$. □

Thus far, we do not have many examples of totally accessible tensor norms. Our next result will broaden our knowledge considerably.

Proposition 7.13. *If α is a left accessible tensor norm then its right injective associate $\alpha\backslash$ is totally accessible.*

Proof. Let X and Y be Banach spaces. Let $C(K)$ contain Y as a subspace. Then $X \otimes_{\alpha\backslash} Y$ is a subspace of $X \otimes_\alpha C(K)$. Furthermore, since α is left accessible and $C(K)$ has the metric approximation property, $X \otimes_\alpha C(K)$ is a subspace of $(X^* \otimes_{\alpha'} C(K)^*)^*$.

On the other hand, $C(K)^*$ is a $\mathcal{L}_{1,\lambda}$-space for every $\lambda > 1$ and so $X^* \otimes_{\alpha'/} Y^*$ is a quotient space of $X^* \otimes_{\alpha'} C(K)^*$. But $X^* \otimes_{\alpha'/} Y^* = X^* \otimes_{\alpha\backslash'} Y^*$. Therefore $X^* \otimes_{\alpha\backslash'} Y^*$ is a quotient of $X^* \otimes_{\alpha'} C(K)^*$ and it follows that $(X^* \otimes_{\alpha\backslash'} Y^*)^*$ is a subspace of $(X^* \otimes_{\alpha'} C(K)^*)^*$.

The following commutative diagram summarizes these facts:

$$\begin{array}{ccc} X \otimes_{\alpha\backslash} Y & \dashrightarrow & (X^* \otimes_{\alpha\backslash'} Y^*)^* \\ \downarrow & & \uparrow \\ X \otimes_\alpha C(K) & \longrightarrow & (X^* \otimes_{\alpha'} C(K)^*)^* \end{array}$$

We have shown that all the solid arrows are isometric embeddings. Therefore the canonical mapping $X \otimes_{\alpha\backslash} Y \to (X^* \otimes_{\alpha\backslash'} Y^*)^*$ is also an isometric embedding. This completes the proof. □

A moment's thought will reveal that there is no corresponding result for the projective associate.

We can now generate many accessible norms:

Corollary 7.14. *Let α be a tensor norm. Then the tensor norms $/\alpha\backslash$, $(\backslash\alpha)\backslash$ and $/(\alpha/)$ are all totally accessible.*

Now g_p, being left projective, is left accessible. Therefore, we have:

Corollary 7.15. *Let $1 \le p \le \infty$. The tensor norms $g_p\backslash$ and $/d_p$ are totally accessible.*

The case $p = 2$ is special since, by Corollary 6.25, d_2 is already left injective. Therefore:

Corollary 7.16. *The tensor norms d_2 and g_2 are totally accessible.*

7.3 The Chevet–Saphar Dual Norms and p-integral Operators

We are now ready to find the duals of the Chevet–Saphar tensor norms. The identification of the dual norms will lead to a new class of operators.

Proposition 7.17. *Let $1 \leq p < \infty$ and let X be a Banach space. Then $\ell_1^n \otimes_{g_p} X = \mathcal{P}_p(\ell_\infty^n, X)$ for every n.*

Proof. Each tensor $u = \sum_{j=1}^m z_j \otimes x_j \in \ell_1^n \otimes Y$ corresponds to the operator $u\colon \ell_\infty^n \to X$ given by $u(y) = \sum_{j=1}^m \langle z_j, y\rangle x_j$. Since ℓ_∞^n is finite dimensional, every operator from this space is p-summing and so the spaces $\ell_1^n \otimes_{g_p} X$ and $\mathcal{P}_p(\ell_\infty^n, X)$ coincide algebraically.

We show first that $\pi_p(u) \leq g_p(u)$. To see this, take a representation u as above. We may assume that $z_j \neq 0$ for every j. Now let (y_k) be a finite sequence in ℓ_∞^n. Then

$$\sum_k \|u(y_k)\|^p = \sum_k \Big\|\sum_j \langle z_j, y_k\rangle x_j\Big\|^p .$$

For each k, let $\varphi_k \in B_{X^*}$ satisfy

$$\Big\|\sum_j \langle z_j, y_k\rangle x_j\Big\| = \varphi_k\Big(\sum_j \langle z_j, y_k\rangle x_j\Big) = \sum_j \langle z_j, y_k\rangle\langle x_j, \varphi_k\rangle .$$

Writing $z_j = \|z_j\|\hat{z}_j$ with $\|\hat{z}_j\| = 1$, we have, with the obvious modifications for the case $p = 1$,

$$\begin{aligned}
\sum_k \|u(y_k)\|^p &= \sum_k \Big(\sum_j \langle z_j, y_k\rangle\langle x_j, \varphi_k\rangle\Big)^p \\
&\leq \sum_k \Big(\sum_j |\langle z_j, y_k\rangle|^p\Big)\Big(\sum_j |\langle x_j, \varphi_k\rangle|^{p'}\Big)^{p/p'} \\
&\leq \sum_k \sum_j |\langle z_j, y_k\rangle|^p \left(\|(x_j)\|_{p'}^w\right)^p \\
&= \sum_j \|z_j\|^p \sum_k |\langle \hat{z}_j, y_k\rangle|^p \left(\|(x_j)\|_{p'}^w\right)^p \\
&\leq \left(\|(z_j)\|_p \,\|(x_j)\|_{p'}^w \,\|(y_k)\|_p^w\right)^p .
\end{aligned}$$

This holds for every representation of u and every finite sequence (y_k) and so, by the definitions of these norms, $\pi_p(u) \leq g_p(u)$.

To prove the reverse inequality, we apply the Pietsch Domination Theorem (Theorem 6.18) to the operator u, taking K to be the set $\{e_1, \ldots, e_n\}$ of unit basis vectors in ℓ_1^n. Therefore there exist nonnegative real numbers $\mu_1, \ldots, \mu_n$ such that $\sum_{i=1}^n \mu_i = 1$ and

$$\|u(y)\|^p \le \pi_p(u)^p \sum_{i=1}^{n} \mu_i |y_i|^p$$

for every $y \in \ell_\infty^n$. For each i, let $z_i = \mu_i^{1/p} e_i$ and $x_i = \beta_i u(e_i)$, where $\beta_i = \mu_i^{-1/p}$ if $\mu_i \neq 0$ and $\beta_i = 0$ if $\mu_i = 0$. Then $u(y) = \sum_{i=1}^n y_i u(e_i) = \sum_{i=1}^n \langle z_i, y\rangle x_i$ for every $y \in \ell_\infty^n$. We estimate $g_p(u)$ using the representation $\sum_{i=1}^n z_i \otimes x_i$: we have $\|(z_i)\|_p = \left(\sum_i \mu_i\right)^{1/p} = 1$ and

$$\begin{aligned} \|(x_i)\|_{p'}^w &= \sup\left\{ \left\| \sum_i \alpha_i x_i \right\| : \sum_i |\alpha_i|^p \le 1 \right\} \\ &= \sup\left\{ \left\| u\big(\sum_i \alpha_i \beta_i e_i\big) \right\| : \sum_i |\alpha_i|^p \le 1 \right\} \\ &\le \pi_p(u) \sup\left\{ \sum_i \mu_i |\alpha_i|^p \beta_i^p : \sum_i |\alpha_i|^p \le 1 \right\} = \pi_p(u)\,. \end{aligned}$$

Therefore $g_p(u) \le \|(z_i)\|_p \|(x_i)\|_{p'}^w \le \pi_p(u)$. □

There is also a dual result in finite dimensions:

Corollary 7.18. *Let $1 \le p < \infty$ and let E be finite dimensional. Then $E \otimes_{g_p} \ell_\infty^n = \left(E^* \otimes_{d_{p'}} \ell_1^n\right)^* = \mathcal{P}_p(E^*, \ell_\infty^n)$ for every n.*

Proof. The transpose $u \mapsto u^t$ is an isometric isomorphism between $X \otimes_{g_p} Y$ and $Y \otimes_{d_p} X$. Hence $E \otimes_{g_p} \ell_\infty^n$ is isometrically isomorphic with $\ell_\infty^n \otimes_{d_p} E$. But

$$\ell_\infty^n \otimes_{d_p} E = \mathcal{P}_{p'}(\ell_\infty^n, E^*)^* = \left(\ell_1^n \otimes_{g_{p'}} E^*\right)^*$$

by the above proposition. Applying the transpose again to the last space in this sequence gives the space $\left(E^* \otimes_{d_{p'}} \ell_1^n\right)^* = \mathcal{P}_p(E^*, \ell_\infty^n)$. Hence $E \otimes_{g_p} \ell_\infty^n = \mathcal{P}_p(E^*, \ell_\infty^n)$. □

At this point, it will be helpful to interchange the roles of the conjugate indices p and p' and view this result as a statement about the dual of the tensor norm d_p in a special case. Thus, we have

$$E \otimes_{d_p'} \ell_\infty^n = \left(E^* \otimes_{d_p} \ell_1^n\right)^* = E \otimes_{g_p} \ell_\infty^n$$

for every n and every finite dimensional space E. We now extend this statement to infinite dimensions. Every $C(K)$ space is a $\mathcal{L}_{\infty,\lambda}$-space for every $\lambda > 1$. Therefore, since d_p' and g_p are tensor norms, we have:

Proposition 7.19. *Let $1 < p \le \infty$ and let X be a Banach space. Then the norms d_p' and g_p coincide on $X \otimes C(K)$ for every K.*

Now the norms d_p are right projective and so the dual norms d'_p are right injective. Proposition 7.9 tells us that the right injective associate norm $g_p\backslash$ on any tensor product $X\otimes Y$ is induced by the embedding $X\otimes Y \to X\otimes_{g_p} C(K)$, where $C(K)$ contains Y as a subspace. Putting these facts together with the above result, we obtain an identification of the dual norm of d_p:

Theorem 7.20. *Let $1 < p \leq \infty$. The dual of the tensor norm d_p is the right injective associate norm $g_{p'}\backslash$.*

Transposition gives a corresponding statement for g_p, namely, $g'_p = /d_{p'}$ and, taking duals, $g_p = (/d_{p'})'$. Pursuing this a little further, we see that

$$g_p = (/d_{p'})' = \backslash(d'_{p'}) = \backslash(g_p\backslash)\,.$$

On the other hand, g_p is left projective, and so $(\backslash g_p)\backslash = g_p\backslash$. Summarizing, we have

$$\backslash(g_p\backslash) = g_p \qquad \text{and} \qquad (\backslash g_p)\backslash = g_p\backslash\,. \tag{7.6}$$

The case $p = 2$ is particularly pleasant. We have seen that g_2 is right injective. Therefore

$$d'_2 = g_2 \qquad \text{and} \qquad g'_2 = d_2\,. \tag{7.7}$$

The identification of the duals of the Chevet–Saphar norms enables us to establish their accessibility. We have shown in Corollary 7.15 that $g_p\backslash$ is totally accessible. Therefore, by Proposition 7.4:

Proposition 7.21. *The tensor norms g_p and d_p are accessible for every p.*

We now address the question of describing the dual space of $X \otimes_{d'_p} Y$ in terms of a space of operators. It will be convenient to work with the space $X \otimes_{g'_p} Y$; the dual space for the norm d'_p can be obtained simply by transposing X and Y.

Let I be the canonical embedding of X into $C(B_{X^*})$. Then, by Proposition 7.9, $X\otimes_{g'_p} Y = X\otimes_{/d_{p'}} Y$ is a subspace of $C(B_{X^*})\otimes_{d_{p'}} Y$. If $T \in \mathcal{L}(X, Y^*)$ is a bounded linear functional on $X \otimes_{g'_p} Y$ then, by the Hahn–Banach Theorem, T may be extended to a bounded linear functional on $C(B_{X^*}) \otimes_{d_{p'}} Y$ with the same norm. Denoting the extension by S, we have the following factorization of T:

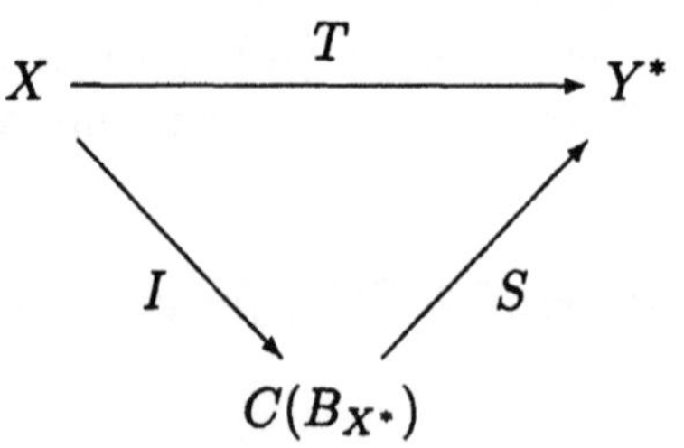

Now S, belonging as it does to the dual space of $C(B_{X^*}) \otimes_{d_{p'}} Y$, is a p-summing operator from $C(B_{X^*})$ into Y^* and the p-summing norm of S is equal to the norm of the linear functional T. By Proposition 6.22 there exists a regular Borel probability measure μ on B_{X^*} and an operator $U\colon L_p(\mu) \to Y^*$ such that $S = UJ_p$ and $\pi_p(S) = \|U\|$. We now have a factorization of T:

$$\begin{array}{ccc} X & \xrightarrow{\quad T \quad} & Y^* \\ \Big\downarrow I & & \Big\uparrow U \\ C(B_{X^*}) & \xrightarrow{\ J_p\ } & L_p(\mu) \end{array}$$

Furthermore, the norm of T as an element of the dual space $(X \otimes_{g'_p} Y)^*$ is equal to $\|U\|$. When $p = 1$ this is simply the description of an integral operator and so the following terminology is appropriate. We shall say that an operator $T\colon X \to Z$ is a *p-integral operator* if the corresponding bilinear form on $X \times Z^*$ defines a bounded linear form on $X \otimes_{g'_p} Z^*$. The *p-integral norm* of T, which we denote by $i_p(T)$ is defined to be the norm of this functional. We denote the space of p-integral operators, equipped with this norm, by $\mathcal{J}_p(X, Z)$.

Theorem 7.22. *Let $1 < p \le \infty$ and let X and Y be Banach spaces. An operator $T\colon X \to Y$ is p-integral if and only if there exists a compact space K, a regular Borel probability measure μ on K and operators $S\colon X \to C(K)$ and $R\colon L_p(\mu) \to Y^{**}$ such that $JT = RI_pS$, where J is the canonical embedding of Y into Y^{**} and I_p is the canonical mapping of $C(K)$ into $L_p(\mu)$. Furthermore, the p-integral norm of T is given by*

$$i_p(T) = \inf \|S\|\|R\|\,,$$

where the infimum is taken over all such factorizations, and this infimum is attained:

$$\begin{array}{ccccc} X & \xrightarrow{\quad T \quad} & Y & \xrightarrow{\quad J \quad} & Y^{**} \\ \Big\downarrow S & & & & \Big\uparrow R \\ C(K) & & \xrightarrow{\qquad I_p \qquad} & & L_p(\mu) \end{array}$$

Proof. Suppose that T is a p-integral operator. Then the argument given above shows that the operator JT has a factorization of the form given in the statement of the proposition, with $i_p(T) = \|R\|\|S\|$.

Conversely, suppose that JT has such a factorization. We may factor the canonical mapping of $C(K)$ into $L_p(\mu)$ through the canonical mapping from $L_\infty(\mu)$ in $L_p(\mu)$, which we also denote by I_p. We now have the following diagram:

$$\begin{array}{ccccc} X & \xrightarrow{T} & Y & \xrightarrow{J} & Y^{**} \\ \Big\downarrow{\scriptstyle S_1} & & & & \Big\uparrow{\scriptstyle R} \\ L_\infty(\mu) & & \xrightarrow{\quad I_p \quad} & & L_p(\mu) \end{array}$$

where $\|S_1\| = \|S\|$. By the injectivity of $L_\infty(\mu)$, the operator S_1 extends to an operator $S_2\colon C(B_{X^*}) \to L_\infty(\mu)$ with the same norm as S. This yields the following factorization of JT:

$$\begin{array}{ccccc} X & \xrightarrow{T} & Y & \xrightarrow{J} & Y^{**} \\ \Big\downarrow{\scriptstyle I} & & & & \Big\uparrow{\scriptstyle R} \\ C(B_{X^*}) & & \xrightarrow{\quad V \quad} & & L_p(\mu) \end{array}$$

where $I\colon X \to C(B_{X^*})$ is the canonical embedding and $V = I_pX_2$. Then V is a p-summing operator with $\pi_p(v) \leq \|S_2\| = \|S\|$. Therefore the operator RV defines a bounded linear functional on $C(B_{X^*}) \otimes_{d_{p'}} Y^*$ with norm at most $\pi_p(RV) \leq \|R\|\pi_p(V) \leq \|R\|\|V\|$. Taking the restriction of this functional to the subspace $X \otimes_{g'_p} Y^*$ of $C(B_{X^*}) \otimes_{d_{p'}} Y^*$, we see that T belongs to the dual space $(X \otimes_{g'_p} Y^*)^*$ and its norm is at most $\|R\|\|S\|$. □

We now have a representation of the dual space of $X \otimes_{g'_p} Y$:

$$(X \otimes_{g'_p} Y)^* = \mathcal{I}_p(X, Y^*)\,.$$

Furthermore, we have an isometric embedding of $\mathcal{I}_p(X,Y)$ into $\mathcal{I}_p(X,Y^{**})$ and this give the canonical embedding:

$$\mathcal{I}_p(X,Y) \subset (X \otimes_{g'_p} Y^*)^*\,.$$

The p-integral operators share many of the properties of the integral (or 1-integral) operators. We summarize some of them in the following proposition. The proof is left to the reader.

Proposition 7.23. *Let $T\colon X \to Y$ be an operator.*

(a) *If J is the canonical embedding of Y into Y^{**} then T is p-integral if and only if either of the operators JT, T^{**} is p-integral. The p-integral norms of T, JT and T^{**} are the same.*

(b) *If T is a p-integral operator then for all operators $R: W \to X$ and $S: Y \to Z$, the operator STR is p-integral and*

$$i_p(STR) \le \|S\| i_p(T) \|R\| .$$

(c) *T is p-integral if and only if there exists a positive constant C such that for every finite dimensional subspace E of X and every finite codimensional subspace F of Y, the finite dimensional operator*

$$Q_F T I_E : E \to X \to Y \to Y/F$$

satisfies $i_p(Q_F T I_E) \le C$. Furthermore, $i_p(T) = \inf C$, where the infimum is taken over all such pairs E, F.

It is interesting to compare the classes of p-integral and p-summing operators. The factorization schemes that characterize these classes tell the whole story. First, for the p-integral operators:

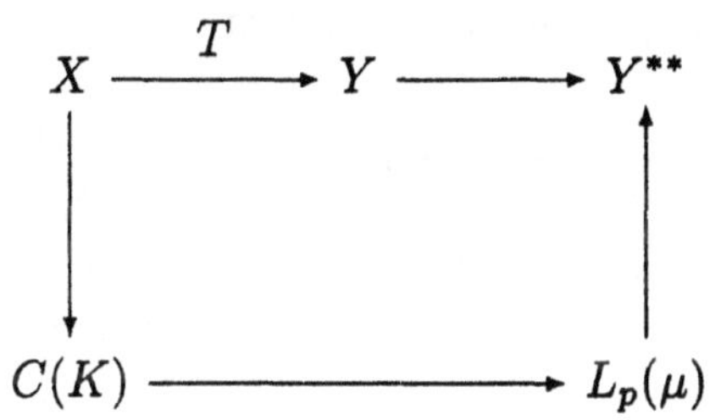

And for the p-summing operators:

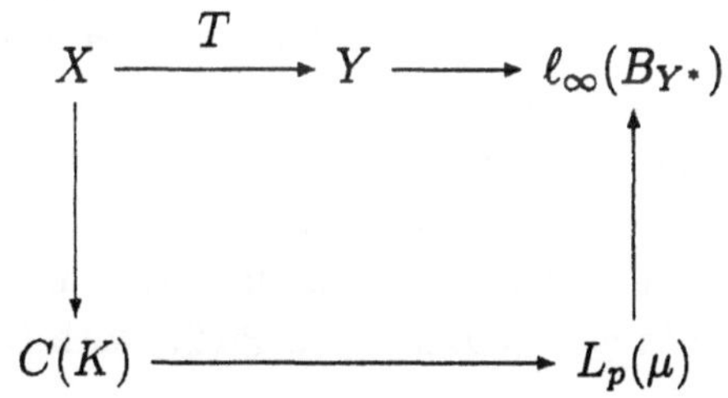

The only difference is that in the second case, the canonical embedding of Y into Y^{**} is replaced by the embedding of Y into the associated injective space $\ell_\infty(B_{Y^*})$ (or any injective space that contains Y). It follows immediately that every p-integral operator is p-summing and that

$$\pi_p(T) \le i_p(T) .$$

In general, these classes are not the same. For example, we have seen that the injection of ℓ_1 into ℓ_2 is 1-summing, but, bearing in mind that ℓ_2 has the Radon–Nikodým property, one can see that this operator is not 1-integral. There are other significant differences between the p-integral and p-summing operators. The p-summing nature of an operator is not altered, nor is its

p-summing norm, if the range space is enlarged. This is clearly not the case for a p-integral operator.

We conclude with an examination of the consequences of the accessibility of the Chevet–Saphar norms.

Proposition 7.24. *Let $1 \leq p \leq \infty$ and let X and Y be Banach spaces. If X^* or Y has the metric approximation property, then the space of p-nuclear operators from X into Y is isometrically isomorphic to $X^* \hat{\otimes}_{g_p} Y$ and the p-nuclear and p-integral norms coincide on this space.*

Proof. By the accessibility of g_p, the canonical mapping from $X^* \otimes_{g_p} Y$ into $(X \otimes_{g'_p} Y^*)^* = \mathcal{I}_p(X, Y^{**})$ is an embedding. Hence this mapping extends to an embedding of the completed tensor product $X^* \hat{\otimes}_{g_p} Y$ into $\mathcal{I}_p(X, Y^{**})$. But each element of $X^* \hat{\otimes}_{g_p} Y$ takes its values in Y. The result now follows from the fact that $\mathcal{I}_p(X, Y)$ is a subspace of $\mathcal{I}_p(X, Y^{**})$. □

7.4 The Hilbertian Tensor Norm

In this section, we find a concrete representation of the biggest injective tensor norm, $/\pi\backslash$. We take the following approach to this problem. By Proposition 7.9, the norm $/\pi\backslash$ is induced on $X \otimes Y$ by embedding X and Y into the projective tensor product of suitable $C(K)$ spaces. For example, we have

$$X \otimes_{/\pi\backslash} Y \subset C(B_{X^*}) \otimes_\pi C(B_{Y^*}).$$

Thus, if we can find an injective tensor norm, α, that is equivalent to the projective norm on tensor products of $C(K)$ spaces, then this norm will be equivalent to $/\pi\backslash$ in general. Furthermore, by the finitely generated nature of tensor norms and the fact that every $C(K)$ space is a $\mathcal{L}_{\infty,\lambda}$-space for every $\lambda > 1$, it will suffice to show that α is equivalent to π on $\ell_\infty^n \otimes \ell_\infty^n$ for every n, provided, of course, the equivalence is uniform in n. In other words, we seek a relation of the form $C\alpha \leq \pi \leq D\alpha$ on $\ell_\infty^n \otimes \ell_\infty^n$, with C and D independent of n.

One of the consequences of Grothendieck's Inequality (Theorem 6.31) states that $\pi \leq K_G\, d_2$ and $\pi \leq K_G\, g_2$ on $\ell_\infty^n \otimes \ell_\infty^n$. However, d_2 is only injective on the left, while g_2 is only right injective. A little hopeful guesswork suggests the following definition. We define the norm w_2 on the tensor product $X \otimes Y$ of any two Banach spaces by

$$w_2(u) = \inf\left\{ \|(x_j)\|_2^w \|(y_j)\|_2^w : u = \sum_{j=1}^n x_j \otimes y_j \right\}.$$

We leave it as an exercise to the reader to verify that w_2 is a tensor norm. For reasons that will soon be apparent, w_2 is known as the *Hilbertian tensor norm*. We now show that this norm satisfies one of our requirements.

Theorem 7.25. *The tensor norms* π *and* w_2 *satisfy* $w_2 \le \pi \le K_G\, w_2$ *on* $\ell_\infty^n \otimes \ell_\infty^n$ *for every* n.

Proof. We prove the real case first. Let $u \in \ell_\infty^n \otimes \ell_\infty^n$. To compute the projective norm of u, let B be a bilinear form on $\ell_\infty^n \times \ell_\infty^n$ of unit norm. Thus $B(s,t) = \sum_{i,j} a_{ij} s_i t_j$, where

$$\sup\left\{ \left| \sum_{i,j} a_{ij} s_i t_j \right| : |s_i|, |t_j| \le 1 \right\} = 1 .$$

Let $\sum_{k=1}^m x_k \otimes y_k$ be a representation of u and for each i, let $u_i = (x_{ki})_k$ and $v_i = (y_{ki})_k$. Then

$$|\langle u, B\rangle| = \left| \sum_k \sum_{i,j} a_{ij} x_{ki} y_{kj} \right| = \left| \sum_{i,j} a_{ij} \sum_k x_{ki} y_{kj} \right| = \left| \sum_{i,j} a_{ij} \langle u_i, v_j \rangle \right|$$

Now $\|u_i\|_2 = \left(\sum_k |x_{ki}|^2\right)^{1/2} = \left(\sum_k |\langle x_k, e_i\rangle|^2\right)^{1/2} \le \|(x_k)\|_2^w$ and similarly, $\|v_i\|_2 \le \|(y_i)\|_2^w$ for each i. Therefore, by Grothendieck's Inequality,

$$|\langle u, B\rangle| \le K_G\, \|(x_k)\|_2^w \|(y_k)\|_2^w .$$

It follows from the definition of w_2 that $|\langle u, B\rangle| \le K_G\, w_2(u)$ and hence

$$\pi(u) \le K_G\, w_2(u) .$$

For the complex case, consider the tensor $u' = \sum_k x_k \otimes \overline{y_k}$, where $\overline{y_k} = (\overline{y_{kj}})$. It is easy to see that u' is well defined and that $\pi(u') = \pi(u)$ and $w_2(u') = w_2(u)$. Finally, imitating the proof for the real case gives $\pi(u') \le K_G\, w_2(u)$, where K_G is now the complex Grothendieck constant. □

From the argument outlined above, we see that:

Corollary 7.26. $w_2 \le \pi \le K_G\, w_2$ *on every* $C(K)$ *space.*

Our next task is to show that w_2 is injective. Since w_2 is finitely generated, it will be enough to consider finite dimensional spaces. If E and F are finite dimensional and $\sum_{k=1}^n x_k \otimes y_k$ is a representation of $u \in E \otimes F$, we may define operators $S\colon E^* \to \ell_2^n$ and $R\colon \ell_2^n \to F$ by $S\varphi = (\varphi x_k)$ and $R\lambda = \sum_{k=1}^n \lambda_k y_k$. Then $\|(x_k)\|_2^w = \|S\|$ and $\|(y_k)\|_2^w = \|R\|$. If we view u as a operator from E^* into F, then u is the composition RS.

On the other hand, suppose the operator $u\colon E^* \to F$ is factored through some ℓ_2^n, so that there exist operators $S\colon E^* \to \ell_2^n$ and $R\colon \ell_2^n \to F$ such that $u = RS$. Then there exist $x_k \in E$ and $y_k \in F$ such that $S\varphi = (\varphi x_k)$ and $R\lambda = \sum_{k=1}^n \lambda_k y_k$ and it follows that $\sum_{k=1}^n x_k \otimes y_k$ is a representation of u. It is easy to see that the space ℓ_2^n can be replaced by any larger Hilbert space without altering the norms of the operators R and S, or the fact that $u = RS$. Thus, we have proved the following striking description of the tensor norm w_2 in finite dimensions.

Proposition 7.27. *Let E and F be finite dimensional Banach spaces and let $u \in E \otimes F$. Then $w_2(u) = \inf \|R\|\|S\|$, where the infimum is taken over all the factorizations of the operator $u\colon E^* \to F$ through a Hilbert space.*

The "Hilbertian" nature of w_2 exemplified by this result enables us to prove its injectivity.

Theorem 7.28. *The tensor norm w_2 is injective.*

Proof. It suffices to prove that w_2 is injective for finite dimensional spaces. Suppose that the Banach spaces E and F are subspaces of the finite dimensional spaces M and N respectively. Then

$$w_2(u; M \otimes N) \leq w_2(u; E \otimes F)$$

for every $u \in E \otimes F$. To prove the reverse inequality, let $R_1 S_1$ be a factorization of u through a finite dimensional Hilbert space H_1, where u is considered as an operator from M^* into N. Let P denote the quotient operator from M^* onto E^*. Consider the projection Q of H_1 onto the orthogonal complement of $S_1(\ker P)$. We may define an operator S from E^* into $Q(H_1)$ by setting $S(P\varphi) = QS_1(\varphi)$. Then $\|S\| \leq \|S_1\|$. Let H be the closure of $S(E^*)$ and let R be the restriction of R_1 to H, so that $\|R\| \leq \|R_1\|$. Then RS is a factorization of u through the Hilbert space H, and $\|R\|\|S\| \leq \|R_1\|\|S_1\|$. It follows that $w_2(u; E \otimes F) \leq w_2(u; M \otimes N)$. □

Combining these facts about the Hilbertian tensor norm, we have:

Theorem 7.29. $w_2 \leq /\pi\backslash \leq K_G\, w_2$.

Taking duals, we get a description of the biggest projective tensor norm:

Corollary 7.30. $K_G^{-1}\, w_2' \leq \backslash\varepsilon/ \leq w_2'$.

It will be of great interest to determine the classes of operators generated by w_2 and w_2'. Although we do not yet have an explicit formula for the norm w_2', we already have enough information at hand to determine the dual space of $X \otimes_{w_2'} Y$. We use the following general principle, which is nothing more than the definition of the property of finite generation of a tensor norm: if α is a tensor norm, then an operator $T \in \mathcal{L}(X, Y^*)$ defines a bounded linear functional on $X \otimes_\alpha Y$ if and only if there exists a constant C such that, for every finite dimensional subspace E of X and every finite dimensional subspace F of Y, the finite dimensional operator $Q_F T I_E\colon E \to F^*$, considered as an element of $(E \otimes_\alpha F)^* = E^* \otimes_{\alpha'} F^*$, satisfies $\alpha'(Q_F T I_E) \leq C$:

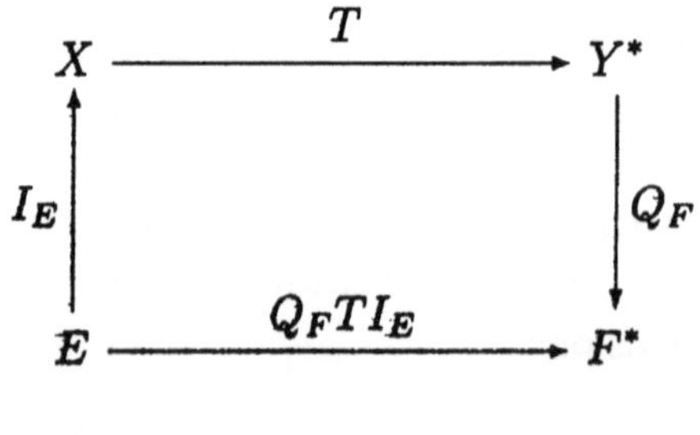

Furthermore, the norm of T in $(X \otimes_\alpha Y)^*$ is the infimum of the set of real numbers C that satisfy this condition.

Thanks to this principle, the proof will reduce to the problem of patching together a family of finite dimensional Hilbertian factorizations of an operator to construct a factorization of the operator itself. For this, we need some topological machinery. Let K be a compact topological space and let $(x_i)_{i\in I}$ be a family of elements of K, indexed by the set I. If $\mathcal{U}$ is an ultrafilter on I, then the limit $\lim_{\mathcal{U}} x_i$ of (x_i) with respect to $\mathcal{U}$ exists in K.

Theorem 7.31. *Let X and Y be Banach spaces. An operator $T\colon X \to Y^*$ defines a bounded linear functional on $X \otimes_{w_2'} Y$ if and only if there exists a Hilbert space H and operators $S\colon X \to H$ and $R\colon H \to Y^*$ such that $T = RS$. Furthermore, the norm of T as an element of $(X \otimes_{w_2'} Y)^*$ is the infimum of $\|R\|\|S\|$, ranging over all such factorizations of T.*

Proof. Suppose that T has such a factorization. Let $u \in X \otimes Y$ and let E, F be finite dimensional subspaces of X, Y respectively such that $u \in E \otimes F$. Since $(E \otimes_{w_2'} F)^* = E^* \otimes_{w_2} F^*$, it follows from Proposition 7.27 that

$$|\langle u, T\rangle| \le \|R\|\|S\| w_2'(u; E \otimes F)\,.$$

Therefore, taking the infimum over E and F, we have

$$|\langle u, T\rangle| \le \|R\|\|S\| w_2'(u; X \otimes Y)\,.$$

Conversely, let $T \in (X \otimes_{w_2'} Y)^*$. Then there exists a constant C such that for every pair, E, F of finite dimensional subspaces of X, Y respectively, there is a factorization $R_{EF}S_{EF}$ of T through a Hilbert space H_{EF}, with $\|R_{EF}\|\|S_{EF}\| \le C$. We may assume that $\|R_{EF}\| \le 1$ and $\|S_{EF}\| \le C$ for every E, F:

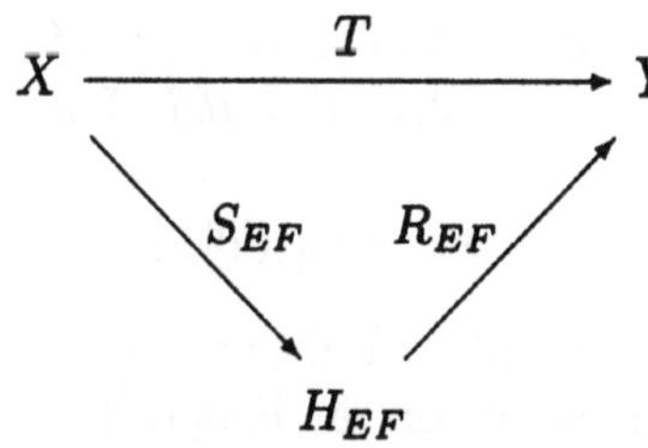

Let H be the ℓ_2-direct sum of the Hilbert spaces H_{EF}. We propose to factor T through a subspace of H. The factoring operators will be defined at each point by taking the limit with respect to a suitable ultrafilter.

We take as indexing set the set I of all pairs (E, F), where E and F are finite dimensional subspaces of X and Y respectively. For each $(E, F) \in I$, let $T(E, F)$ be the set of all pairs (M, N) for which $E \subset M$ and $F \subset N$. These sets form a base for a filter; let $\mathcal{U}$ be any ultrafilter on I that contains this filter.

Fixing $x \in X$, consider the family in H defined by

$$v_{EF} = \begin{cases} S_{EF}x & \text{if} \quad x \in E, \\ 0 & \text{if} \quad x \notin E. \end{cases}$$

By the uniform boundedness of the operators S_{EF}, the family (v_{EF}) lies in a weakly compact subset of H and so it has a weak limit with respect to the ultrafilter $\mathcal{U}$. We define Sx to be this limit. We leave it as a routine exercise to the reader to verify that S is bounded, with $\|S\| \leq C$, and that $\Pi_{EF}Sx = S_{EF}x$ whenever $x \in E$, where Π_{EF} denotes the orthogonal projection of H onto H_{EF}.

We now define an operator R from the subspace $S(X)$ of H into Y^*. Let $v \in S(X)$, so that v is the weak limit of a family (v_{EF}) as defined above, for some $x \in X$. Then the family (Rv_{EF}) is bounded, and so we may define Rv to be its weak* limit in Y^* with respect to $\mathcal{U}$. Once again, we leave it to the reader to verify that R is a well defined operator with $\|R\| \leq 1$, and that $RSx = Tx$ for every $x \in X$. The factorization is completed by taking H_0 to be the closure of $S(X)$ in H and extending R uniquely to H_0:

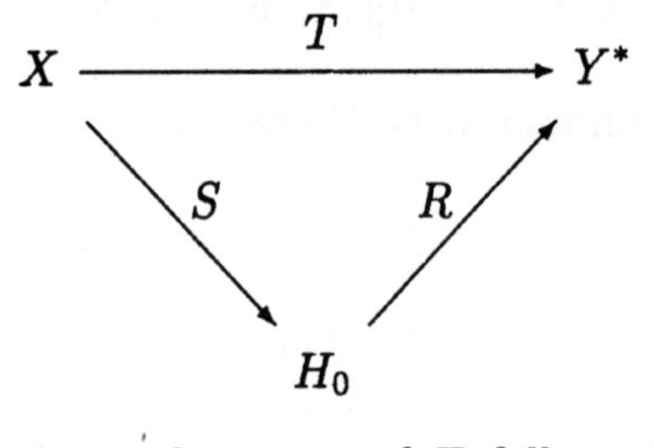

Finally, the assertion about the norm of T follows immediately from the proof. □

An operator $T\colon X \to Y$ is said to be a *Hilbertian operator* (or a *2-factorable operator*) if there exists a Hilbert space H, together with operators $S\colon X \to H$ and $R\colon H \to Y$ such that $T = RS$. The *Hilbertian norm* of T is defined by

$$\|T\|_{\mathcal{H}} = \inf \|R\|\|S\|,$$

the infimum being taken over all such factorizations of T. The observations made about the p-integral operators in Proposition 7.23 also apply to the Hilbertian operators, with the p-integral norm being replaced by the Hilbertian norm. In particular, we note that an operator $T\colon X \to Y$ is Hilbertian if and only if the operator $JT\colon X \to Y^{**}$ is Hilbertian and, furthermore, the Hilbertian norms of T and JT are the same. We denote the Banach space of Hilbertian operators from X into Y, with the Hilbertian norm, by $\mathcal{L}_{\mathcal{H}}(X, Y)$.

We may also define the class of Hilbertian bilinear forms. We say that a bilinear form $B \in \mathcal{B}(X \times Y)$ is a *Hilbertian bilinear form* if the corresponding operator $T\colon X \to Y^*$ is Hilbertian, and we define the Hilbertian norm, $\|B\|_{\mathcal{H}}$

On the other hand, we may also associate an operator $S\colon Y \to X^*$ with B, satisfying $B(x,y) = \langle x, Sy\rangle$. The canonical right extension of B is then defined by

$$B^*(x^{**}, y^{**}) = \langle S^*x^{**}, y^{**}\rangle .$$

It is not difficult to see that ${}^*B = B^*$ on $X^{**} \times Y$ and on $X \times Y^{**}$. Exercise 6.2 gives another viewpoint on these extension processes.

These observations are helpful in analysing the operator versions of the α-integral bilinear forms. Let $T\colon X \to Y$ be an operator and let B be the corresponding bilinear form on $X \times Y^*$, so that

$$B(x,\psi) = \langle Tx, \psi\rangle$$

for every $x \in X$ and $\psi \in Y^*$. The operator from X into Y^{**} that corresponds to this bilinear form is $J_Y T$, where J_Y is the canonical embedding of Y into Y^{**}. Conversely, starting with a bilinear form B on $X \times Y$, we obtain an operator T from X into Y^* which in turn is associated with the bilinear form ${}^*B = B^*$ on $X \times Y^{**}$. We shall say that the operator $T\colon X \to Y$ is an *α-integral operator* if the associated bilinear form B on $X \times Y^*$ is an α-integral bilinear form. The *α-integral norm* of T, denoted by $\|T\|_\alpha$, is defined to be the α-integral norm of B. The facts outlined above prove the following result.

Proposition 8.2. *The following are equivalent for an operator $T\colon X \to Y$:*

(i) *T is an α-integral operator.*
(ii) *$J_Y T\colon X \to T^{**}$ is an α-integral operator.*
(iii) *T^{**} is an α-integral operator.*

Furthermore, the α-integral norms of T, $J_Y T$ and T^{**} are the same.

In particular, we have the canonical embeddings

$$\mathcal{L}_\alpha(X,Y) \subset (X \otimes_{\alpha'} Y^*)^* \subset \mathcal{L}_\alpha(X, Y^{**}) .$$

The examples considered above translate immediately into the setting of operators. Thus, the α-integral operators in the cases $\alpha = \varepsilon$ and π are the integral and the bounded operators respectively. The p-integral and p-summing operators are also α-integral:

$$\mathcal{L}_{g_p}(X,Y) = \mathcal{I}_p(X,Y)$$

and

$$\mathcal{L}_{g_p\backslash}(X,Y) = \mathcal{P}_p(X,Y) .$$

And, of course, our previous identifications of the classes of w_2- and w_2'-integral bilinear forms may also be couched in terms of operators:

$$\mathcal{L}_{w_2}(X,Y) = \mathcal{L}_{\mathcal{H}}(X,Y)$$

and

$$\mathcal{L}_{w_2'}(X,Y) = \mathcal{D}_2(X,Y).$$

Each property of α-integral bilinear forms has a counterpart for the α-integral operators. We leave the proof of the following proposition as an exercise.

Proposition 8.3. *Let α be a tensor norm.*

(a) *An operator $T: X \to Y$ is α-integral if and only if there exists a constant C such that, for every finite dimensional subspace E of X and every finite codimensional subspace F of Y, the finite dimensional operator $Q_F T I_E: E \to Y/F$ satisfies*

$$\|Q_F T I_E\|_\alpha \leq C.$$

(b) *Let $T: X \to Y$ be an α-integral operator and let $S: W \to X$ and $R: Y \to Z$ be operators. Then the operator RTS is α-integral and*

$$\|RTS\|_\alpha \leq \|R\|\|T\|_\alpha\|S\|.$$

The following result provides a useful test for an operator to be α-integral.

Theorem 8.4. *An operator $T: X \to Y$ is α-integral if and only if the tensor product operator $T \otimes I: X \otimes_{\alpha'} Y^* \to Y \otimes_\pi Y^*$ is bounded, where I is the identity operator on Y^*. Furthermore, the α-integral norm of T satisfies*

$$\|T\|_\alpha = \|T \otimes I\| = \sup\{|\operatorname{tr}((T \otimes I)u)| : \alpha'(u) \leq 1\}.$$

Proof. Let B be the associated bilinear form on $X \times Y^*$. Then, for $u = \sum_{j=1}^n x_j \otimes \psi_j \in X \otimes Y^*$, we have

$$\langle u, B\rangle = \sum_{j=1}^n \langle Tx_j, \psi_j\rangle = \operatorname{tr}\Big(\sum_{j=1}^n Tx_j \otimes \psi_j\Big)$$

where tr is the trace functional on $Y \otimes_\pi Y^*$. Therefore

$$\langle u, B\rangle = \operatorname{tr}\big((T \otimes I)u\big)$$

for every $u \in X \otimes Y^*$.

It follows that, if $T \otimes I$ is bounded, then

$$|\langle u, B\rangle| \leq \big|\operatorname{tr}((T \otimes I)u)\big| \leq \pi((T \otimes I)u) \leq \|T \otimes I\|\alpha'(u),$$

and so T is α-integral, with

$$\|T\|_\alpha \leq \sup\{|\operatorname{tr}((T \otimes I)u)| : \alpha'(u) \leq 1\} \leq \|T \otimes I\|.$$

We now consider the spaces of "nuclear" forms and operators associated with a tensor norm. Proceeding by analogy with the case $\alpha = \pi$, we begin with the canonical mapping of $X^* \otimes Y^*$ into $\mathcal{B}_\alpha(X \times Y)$. If we endow $X^* \otimes Y^*$ with the tensor norm α, then this mapping is bounded, with norm at most one. Hence, it extends to an operator

$$J_\alpha \colon X^* \hat{\otimes}_\alpha Y^* \to \mathcal{B}_\alpha(X \times Y).$$

The bilinear forms that lie in the range of this operator are known as *α-nuclear bilinear forms.* The *α-nuclear norm* of an α-nuclear bilinear form B is defined by

$$N_\alpha(B) = \inf\{\alpha(u) : u \in X^* \hat{\otimes}_\alpha Y^*,\ J_\alpha u = B\}.$$

We shall denote the space of α-nuclear bilinear forms, with the α-nuclear norm, by $\mathcal{N}_\alpha(X \times Y)$. It is not difficult to see that this space is complete; indeed $\mathcal{N}_\alpha(X \times Y)$ is a quotient space of $X^* \hat{\otimes}_\alpha Y^*$:

$$\mathcal{N}_\alpha(X \times Y) = X^* \hat{\otimes}_\alpha Y^* / \ker J_\alpha .$$

In the same way, there is a canonical mapping, which we also denote by J_α, from $X^* \hat{\otimes}_\alpha Y$ into $\mathcal{L}_\alpha(X, Y)$. The operators that lie in the range of this mapping are known as *α-nuclear operators.* The α-nuclear norm, $N_\alpha(T)$, of an α-nuclear operator is defined in the same way as for bilinear forms. Again, the space $\mathcal{N}_\alpha(X, Y)$ of α-nuclear operators is the quotient space of $X^* \hat{\otimes}_\alpha Y$ by the kernel of J_α.

We have already encountered several examples of this construction. In the case $\alpha = \pi$, the α-nuclear operators are just the nuclear operators. When α is the injective norm, we obtain the approximable operators. And if α is the Chevet–Saphar norm g_p, the corresponding class consists of the p-nuclear operators. We have seen in Chapter 4 that $\mathcal{N}(X, Y) = X^* \hat{\otimes}_\pi Y$ if either X^* or Y has the approximation property. We now show that this is a special case of a general phenomenon.

Proposition 8.7. *Let α be a tensor norm. If either X^* or Y has the approximation property, then $\mathcal{N}_\alpha(X, Y) = X^* \hat{\otimes}_\alpha Y$.*

Proof. We give the proof for the case in which Y has the approximation property. Supppose that $J_\alpha u = 0$ for some $u \in X^* \hat{\otimes}_\alpha Y$. In order to establish that $u = 0$, we must show that $\langle u, T \rangle = 0$ for every $T \in (X^* \hat{\otimes}_\alpha Y)^* = \mathcal{L}_{\alpha'}(X^*, Y^*)$. By Theorem 8.4, the operator $T \otimes I \colon X^* \hat{\otimes}_\alpha Y \to Y^* \hat{\otimes}_\pi Y$ is bounded. If Y has the approximation property, then $Y^* \hat{\otimes}_\pi Y$ is the space of nuclear operators on Y. Now $J_\alpha u = 0$ is easily seen to imply that the nuclear operator $(T \otimes I)(u)$ is zero. Therefore, the trace functional on $Y^* \hat{\otimes}_\pi Y$ vanishes on $(T \otimes I)(u)$. Hence, by the proof of Theorem 8.4, we have $\langle u, T \rangle = \operatorname{tr}((T \otimes I)(u)) = 0$. □

The question of when $\mathcal{N}_\alpha(X,Y)$ is a closed subspace of $\mathcal{L}_\alpha(X,Y)$ has already been answered. From the discussion at the beginning of this section, we have:

Proposition 8.8. *Let α be a tensor norm and let X and Y be Banach spaces. The α-integral norm coincides with the α-nuclear norm on $\mathcal{N}_\alpha(X,Y)$ if one of the following conditions is satisfied:*

(a) *Both X^* and Y have the metric approximation property;*
(b) *α is accessible and either X^* or Y has the metric approximation property;*
(c) *α is totally accessible.*

In finite dimensions, the classes of α-nuclear and α-integral operators are the same. In a sense that will be made clear in the next section, the α-nuclear and α-integral operators respectively constitute the smallest and biggest extension of this class of operators to infinite dimensions.

8.2 Operator Ideals

The theory of Banach operator ideals provides another means to study classes of operators on Banach spaces, without explicit reference to tensor norms. In this section, we give a brief introduction to Banach ideals.

A *Banach ideal* consists of an assignment to each pair of Banach spaces X, Y of a vector space $\mathcal{A}(X,Y)$ of bounded operators from X into Y, together with a norm, A, on this space, with the following properties:

(I1) The one-dimensional operator $\varphi\otimes y$ belongs to $\mathcal{A}(X,Y)$ for every $\varphi \in X^*$ and $y \in Y$, and $A(\varphi \otimes y) \leq \|\varphi\|\|y\|$.
(I2) If $S \in \mathcal{A}(X,Y)$ and if $T \in \mathcal{L}(W,X)$ and $R \in \mathcal{L}(Y,Z)$, then the composition RST belongs to $\mathcal{A}(W,Z)$ and $A(RST) \leq \|R\|A(S)\|T\|$.
(I3) $[\mathcal{A}(X,Y), A]$ is a Banach space.

We shall use the notation $[\mathcal{A}, A]$ to refer to a Banach ideal, or simply $\mathcal{A}$ if the norm is understood. Almost all the classes of operators we have encountered constitute Banach ideals. Clearly, the largest Banach ideal consists of all the bounded operators on Banach spaces, with the operator norm. At the other end of the scale, we have the ideal $[\mathcal{N}, N]$ of nuclear operators: for each pair of Banach spaces X, Y, $\mathcal{N}(X,Y)$ is the space of nuclear operators and $N(T)$ is the nuclear norm of T. Between these extremes, we have the various Banach ideals of p-nuclear, p-integral and p-summing operators, and the ideals of Hilbertian and 2-dominated operators. In addition to these, the classes of approximable, compact, weakly compact and completely continuous operators, and so on, each define a Banach ideal, where the ideal norm in each case is the operator norm.

Every tensor norm generates two Banach ideals. Given a tensor norm α, we have on the one hand the ideal $[\mathcal{N}_\alpha, N_\alpha]$, where $\mathcal{N}_\alpha(X,Y)$ is the space of α-nuclear operators and N_α the α-nuclear norm. On the other hand, we may form the ideal $[\mathcal{L}_\alpha, \|\cdot\|_\alpha]$, consisting of the α-integral operators with the α-integral norm.

To go in the opposite direction, suppose $[\mathcal{A}, A]$ is a Banach ideal. We may generate a tensor norm in the following way. First, if E and F are finite dimensional spaces, consider the space $\mathcal{A}(E^*, F)$. It follows from property (I1) that this space contains all the operators from E^* into F. Thus $\mathcal{A}(E^*, F) = E \otimes F$ and so we may define a norm on the tensor product $E \otimes F$ by $\alpha(u) = A(u)$. Thus

$$E \otimes_\alpha F = [\mathcal{A}(E^*, F), A] \tag{8.1}$$

or, equivalently,

$$[\mathcal{A}(E, F), A] = E^* \otimes_\alpha F\,. \tag{8.2}$$

It is a simple matter to show that properties (I1) and (I2) imply that α is a uniform crossnorm on the class of finite dimensional spaces. Indeed, finite dimensional Banach ideals are just finite dimensional uniform crossnorms in disguise. The passage to infinite dimensions is more delicate. We may apply the process of finite generation to obtain a tensor norm:

$$\alpha(u) = \inf\{\alpha(u; E \otimes F) : E \subset X, F \subset Y, u \in E \otimes F\}$$

where u belongs to the tensor product of the Banach spaces X, Y and the infimum extends over all pairs of finite dimensional subspaces E, F for which $u \in E \otimes F$. We shall refer to α as the *tensor norm associated with the Banach ideal* $\mathcal{A}$. The basic properties of the associated tensor norm are as follows.

Theorem 8.9. *Let* $[\mathcal{A}, A]$ *be a Banach ideal.*

(a) *There is a unique tensor norm* α *such that*

$$[\mathcal{A}(E, F), A] = E^* \otimes_\alpha F$$

for every pair of finite dimensional spaces E, F.

(b) *For every pair of Banach spaces* X, Y, *every* α*-nuclear operator* T *from* X *into* Y *belongs to* $\mathcal{A}(X, Y)$ *and furthermore,*

$$A(T) \leq N_\alpha(T)\,.$$

(c) *For every pair of Banach spaces* X, Y, *every operator* T *in* $\mathcal{A}(X, Y)$ *is an* α*-integral operator and furthermore,*

$$\|T\|_\alpha \leq A(T)\,.$$

Proof. We have already proved (a). To establish (b), let $T \in X^* \otimes Y$, so that T is a finite rank operator from X into Y. Let M be a finite codimensional subspace of X and N a finite dimensional subspace of Y such that

$T \in (X/M)^* \otimes N$. Then the operator T factors through a finite dimensional operator $T_{MN}: X/M \to N$ as shown in the diagram below:

$$\begin{array}{ccc} X & \xrightarrow{\quad T \quad} & Y \\ {\scriptstyle Q_M}\big\downarrow & & \big\uparrow{\scriptstyle I_N} \\ X/M & \xrightarrow{\quad T_{MN} \quad} & N \end{array}$$

Since the quotient operator Q_M and the embedding I_N each have unit norm, it follows from property (I2) that

$$A(T) \leq A(T_{MN}) = \alpha(T; (X/M)^* \otimes N),$$

by the definition of α. Taking the infimum over M and N, we get

$$A(T) \leq \alpha(T)$$

for every $T \in X^* \otimes Y$. It follows immediately from the definition of the α-nuclear norm that every α-nuclear operator T belongs to $\mathcal{A}(X,Y)$, with $A(T) \leq N_\alpha(T)$.

To prove (c), let $T \in \mathcal{A}(X,Y)$. Let E be a finite dimensional subspace of X and let F be a finite codimensional subspace of Y. Now let T_{EF} denote the composition $Q_F T I_E$, as shown in the diagram below:

$$\begin{array}{ccc} E & \xrightarrow{\quad T_{EF} \quad} & Y/F \\ {\scriptstyle I_E}\big\downarrow & & \big\uparrow{\scriptstyle Q_F} \\ X & \xrightarrow{\quad T \quad} & Y \end{array}$$

By property (I2), we have $A(T_{EF}) \leq A(T)$. Now $T_{EF} \in E^* \otimes (Y/F)$ and $A(T_{EF}) = \alpha(T_{EF}) = \|T_{EF}\|_\alpha$. Therefore $\|T_{EF}\|_\alpha \leq A(T)$ for all E, F. Taking the supremum and applying Proposition 8.3 yields the desired conclusion.

It is clear from this theorem that the ideals of integral and nuclear operators associated with a tensor norm are the "maximal" and "minimal" Banach ideals respectively. We would like to develop this idea a little further. In particular, it would be helpful to have intrinsic characterizations of these ideals. A workable duality theory of Banach ideals is essential to achieve this objective.

In finite dimensions we have trace duality: if E and F are finite dimensional spaces, then the spaces of operators $\mathcal{L}(E,F)$ and $\mathcal{L}(F,E)$ are in duality. If $T \in \mathcal{L}(E,F)$ and $S \in \mathcal{L}(F,E)$, we have

$$\langle S, T \rangle = \mathrm{tr}_E\, ST = \mathrm{tr}_F\, TS\,. \tag{8.3}$$

Now, if $[\mathcal{A}, A]$ is a Banach ideal, we may use trace duality to define a dual norm for the ideal norm A; If S is an operator from E into F, we define

$$A^*(S) = \sup\{|\langle S, T\rangle| : T \in \mathcal{L}(F, E),\ A(T) \le 1\}\,. \tag{8.4}$$

As we have seen in Chapter 1, this is essentially the same as the tensor duality between $E^* \otimes F$ and $E \otimes F^*$. The above operators S, T correspond to tensors $u \in E^* \otimes F$ and $v \in E \otimes F^*$. Now, if α is the tensor norm associated with a Banach ideal $[\mathcal{A}, A]$, then

$$(\mathcal{A}(E, F))^* = (E^* \otimes_\alpha F)^* = E \otimes_{\alpha'} F^* = \mathcal{L}_{\breve{\alpha}}(F, E)\,. \tag{8.5}$$

We now proceed exactly as with the definition of the dual of a tensor norm, using finite generation to define the dual ideal. Thus, for each pair X, Y of Banach spaces and each operator $T\colon X \to Y$, we define

$$A^*(T) = \sup\{A^*(Q_F T I_E) : \dim E, \dim Y/F < \infty\}\,, \tag{8.6}$$

the supremum being taken over all pairs E, F, where E is a finite dimensional subspace of X and F is a finite codimensional subspace of Y:

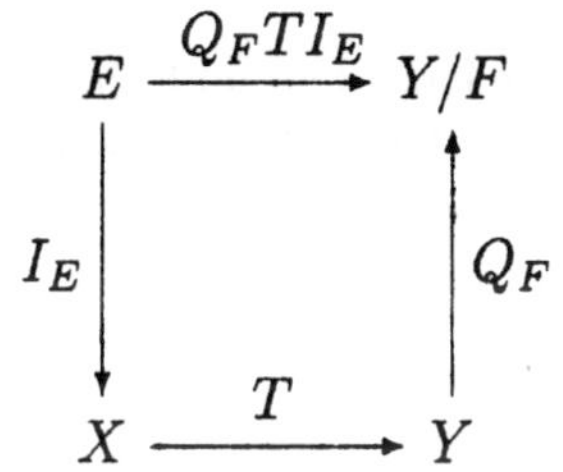

We define $\mathcal{A}^*(X, Y)$ to be the set of operators T for which $A^*(T)$ is finite.

A careful study of this process reveals that $\mathcal{A}^*(X, Y)$ is precisely the space $\mathcal{L}_{\breve{\alpha}}(X, Y)$ of $\breve{\alpha}$-integral operators and A^* is the $\breve{\alpha}$-integral norm. In particular, this shows that $[\mathcal{A}^*, A^*]$ is indeed a Banach ideal. This ideal is known as the *adjoint ideal* of $[\mathcal{A}, A]$. We summarize our findings:

Proposition 8.10. *Let $[\mathcal{A}, A]$ be a Banach ideal and let α be the associated tensor norm. The adjoint ideal $[\mathcal{A}^*, A^*]$ coincides with the Banach ideal of $\breve{\alpha}$-integral operators and the adjoint norm A^* is the $\breve{\alpha}$-integral norm.*

We define a partial ordering on the class of Banach ideals as follows: $[\mathcal{A}, A] \le [\mathcal{B}, B]$ means that $\mathcal{A}(X, Y)$ is contained in $\mathcal{B}(X, Y)$ for all Banach spaces X and Y and $B(T) \le A(T)$ for every $T \in \mathcal{A}(X, Y)$.

We may now define a Banach ideal $[\mathcal{A}, A]$ to be *maximal* if the only Banach ideal $[\mathcal{B}, B]$ that satisfies $[\mathcal{A}, A] \le [\mathcal{B}, B]$ and $B(T) = A(T)$ for every finite rank operator is $[\mathcal{A}, A]$ itself. Similarly, $[\mathcal{A}, A]$ is *minimal* if the only Banach ideal $[\mathcal{B}, B]$ for which $[\mathcal{B}, B] \le [\mathcal{A}, A]$ and $B(T) = A(T)$ for every finite rank operator is $[\mathcal{A}, A]$.

Theorem 8.9 demonstrates that the minimal ideals are precisely the ideals of α-nuclear operators, where α is a tensor norm. We concentrate our attention on the maximal ideals. These are the exact counterpart of the tensor norms. Let us make a definition that will help us to express this idea. We shall say that a Banach ideal $[\mathcal{A}, A]$ is *finitely generated* if, for every pair of Banach spaces X, Y and every $T \in \mathcal{A}(X,Y)$, we have

$$A(T) = \sup\{A(Q_F T I_E) : \dim E, \dim Y/F < \infty\},$$

where the supremum is taken, as usual, over all pairs E, F, where E is a finite dimensional subspace of X and F is a finite codimensional subspace of Y. This is usually taken as the definition of maximality. However, in the light of our final theorem, there is no danger of confusion:

Theorem 8.11. *The following are equivalent for a Banach ideal* $[\mathcal{A}, A]$*:*

(i) $[\mathcal{A}, A]$ *is maximal.*
(ii) $[\mathcal{A}^{**}, A^{**}] = [\mathcal{A}, A]$.
(iii) *There exists a Banach ideal* $[\mathcal{B}, B]$ *such that* $[\mathcal{A}, A] = [\mathcal{B}^*, B^*]$.
(iv) $[\mathcal{A}, A]$ *is finitely generated.*
(v) $[\mathcal{A}, A]$ *is the ideal of α-integral operators, where α is the tensor norm associated with* $[\mathcal{A}, A]$.

Proof. (i) implies (ii) follows from the fact that, if E and F are finite dimensional, then $\mathcal{A}^{**}(E,F) = \mathcal{A}(E,F)$ and the norms A^{**} and A are the same on these spaces.

(ii) implies (iii) is trivial.

(iii) implies (iv) follows immediately from the definition of the adjoint ideal.

(iv) implies (v) follows from Proposition 8.3.

(v) implies (i) also follows from Proposition 8.3. □

8.3 Exercises

Exercise 8.1. Show that an operator T is α^t-integral if and only if the adjoint operator T^* is α-integral.

Exercise 8.2. Let α be a tensor norm. Show that a bilinear form B on $X \times Y$ is $\backslash\alpha/$-integral if and only if the following is true: for every pair of embeddings $S\colon X \to C(K_1)$ and $T\colon Y \to C(K_2)$, there exists an α-integral bilinear form A on $C(K_1) \times C(K_2)$ such that $B = A \circ (S \times T)$. Show that the $\backslash\alpha/$-integral norm of B is the infimum of $\|A\|_\alpha$.

Exercise 8.3. Formulate and prove a characterization of the $/\alpha\backslash$-integral bilinear forms analogous to that given in the previous exercise for $\backslash\alpha/$-integral forms.

Exercise 8.4. Let α be a tensor norm. Show that an operator $T\colon X \to Y$ is $\alpha\backslash$-integral if and only if for every (respectively, some) embedding U of Y into a $C(K)$ space, the composition UT is an α-integral operator.

Exercise 8.5. Formulate and prove characterizations of the $\alpha/$-integral, $/\alpha$-integral and $\backslash\alpha$-integral operators analogous to that given in the previous exercise.

Exercise 8.6. Show that there is no norm under which the finite rank operators form a Banach ideal.

Exercise 8.7. Let $[\mathcal{A}, A]$ and $[\mathcal{B}, B]$ be Banach ideals and suppose that $\mathcal{A}(X,Y)$ is contained in $\mathcal{B}(X,Y)$ for all Banach spaces X, Y. Show that there exists a constant C such that $B(T) \leq CA(T)$ for every X and Y and every $T \in \mathcal{A}(X,Y)$.

Exercise 8.8. Let $[\mathcal{A}, A]$ and $[\mathcal{B}, B]$ be Banach ideals. For each pair of Banach spaces X, Y, let $\mathcal{A} \circ \mathcal{B}(X,Y)$ consist of all operators $T\colon X \to Y$ for which there exist a Banach space Z and operators $S \in \mathcal{B}(X,Z)$ and $R \in \mathcal{A}(Z,Y)$ such that $T = RS$. Show that $\mathcal{A} \circ \mathcal{B}(X,Y)$ is a vector space. For each $T \in \mathcal{A} \circ \mathcal{B}(X,Y)$, let $A \circ B(T) = \inf A(R)B(S)$, where the infimum is taken over all factorizations of T as described above. Show that $A \circ B$ is a quasinorm on $\mathcal{A} \circ \mathcal{B}(X,Y)$ (a quasinorm q on a vector space W has all the properties of a norm, except that the triangle inequality is replaced by $q(u+v) \leq C\big(q(u) + q(v)\big)$, where $C \geq 1$ is a constant).

Exercise 8.9. Let $[\mathcal{A}, A]$ be a Banach ideal. The *conjugate ideal*, $[\mathcal{A}^{\Delta}, A^{\Delta}]$, is defined as follows: an operator $T \in \mathcal{L}(X,Y)$ belongs to $\mathcal{A}^{\Delta}(X,Y)$ if there exists a constant C such that for every finite rank operator $S\colon Y \to X$, we have $|\operatorname{tr} ST| \leq CA(T)$. A norm is defined for these operators by taking $A^{\Delta}(T)$ to be the infimum of C. Show that $A^*(T) \leq A^{\Delta}(T)$ for every $T \in \mathcal{A}^{\Delta}(X,Y)$, with equality if X and Y both have the metric approximation property.

Exercise 8.10. Show that the ideal of compact operators is neither minimal nor maximal.

Exercise 8.11. Let X, Y and Z be Banach spaces. Show that there is an operator from $(X \otimes_{d_2} Y^*) \otimes_{\pi} (Y \otimes_{d_2} Z)$ into $X \otimes_{\pi} Z$ with the property that $(x \otimes \varphi) \otimes (y \otimes z) \mapsto \varphi(y) x \otimes z$.

A. Suggestions for Further Reading

In this appendix, we present some suggestions for further reading. Our choice is highly personal; the reader may consult the books cited below for a fuller set of references.

The first book dealing with normed tensor products was the 1950 monograph of R. Schatten, "A Theory of Cross-Spaces" [46], which summarized the state of knowledge at that time. The ground-breaking works of A. Grothendieck, "Produits Tensoriels Topologiques et Espaces Nucléaires" [21], appearing in 1955, followed by "Résumé de la Théorie Métrique des Produits Tensoriels Topologiques" [19, 22] and a supplementary paper on the Dvoretzky–Rogers Theorem [20, 23] in 1956, introduced a new cycle of ideas, the ramifications of which are evident in Banach space theory today. The fundamental importance of finite dimensional, or "local", considerations, the classes of integral and summing operators, Grothendieck's Inequality, the approximation property, the central role played by vector measures and the Radon–Nikodým property, to name but a few, are all to be found, in one form or another, in his work. However, these developments were not fully appreciated at the time and virtually the only response for over a dozen years was the paper of Amemiya and Shiga [1] in 1957, which explained some of the results in the first part of the Résumé and provided some of the proofs that had been omitted by Grothendieck. It was not until 1968 that J. Lindenstrauss and A. Pełczynski, in the paper "Absolutely Summing Operators in $\mathcal{L}_p$-Spaces and Applications" [34], brought some of the results in Grothendieck's Résumé to the attention of a wider audience. Most importantly, they formulated Grothendieck's Inequality in the terms in which it is usually presented today. However, the explanation was couched in terms of operators rather than tensor products. In the following year, the paper "The $\mathcal{L}_p$-Spaces" [35], by Lindenstrauss and Rosenthal, appeared. This paper demonstrated the importance of local techniques and introduced the all-important Principle of Local Reflexivity, which had, in a sense, been implicit in the Résumé. These papers led to an upsurge of activity, but now many researchers found it easier to carry out their work using the parallel language of operator ideals. Notable exceptions were the works of Chevet [3, 4], Saphar [44, 43, 45] and Lapresté [31, 32], giving detailed treatments of the tensor norms related to the classes of p-summing, p-integral and p-nuclear operators; the 1970 paper by Kwapien and Pełczynski [30] on unconditional structure in matrix spaces; the paper of Lewis in 1973 [33] on weak compactness in injective tensor products; the solution in 1973 of the approximation problem by Enflo [13] and the construction by Pisier [41] in 1983 of a Banach space P for which the injective and projective tensor products $P \otimes_\varepsilon P$ and $P \otimes_\pi P$ are isomorphic. Also noteworthy is the 1973 paper on Banach operator ideals by Gordon, Lewis and Retherford [18] in which tensor product methods play a crucial role.

The excellent survey by Gilbert and Leih [17] develops the material of the Résumé in tandem with the theory of operator ideals. A. Pietsch's book, "Operator Ideals" [40], published ten years after the Lindenstrauss–Pełczynsky paper, describes the state of the art at that time and contains much of the Grothendieck theory, phrased in operator ideal terms. The publication in 1993 of the encyclopaedic monograph "Tensor Norms and Operator Ideals", by A. Defant and K. Floret [6] draws the two strands of tensor products and operator ideals together and shows their interconnections clearly.

We now make some suggestions to supplement particular chapters.

Chapter 2. Schatten's monograph [46] is worth reading to see the beginnings of the theory. The book by Jameson [28] on nuclear and summing norms is highly recommended. For the Bochner integral, see Diestel–Uhl [10].

Chapter 3. Further details on the Pettis integral can be found in the books by Diestel–Uhl [10] and Talagrand [48].

Chapter 4. The approximation property was introduced by Grothendieck. See the Memoir [21] and the Résumé [19, 22]. Enflo [13] provided the first example of a Banach space without the approximation property.

Chapter 5. The outstanding reference on vector measures and the Radon–Nikodým property is the book of Diestel and Uhl [10] and the supplementary material in [11]. See also the books of Bourgin [2] and van Dulst [49]. We have followed the treatment of Diestel and Uhl closely in this chapter. The papers of Gil de la Madrid [15, 16] motivated much of our treatment of tensor products with spaces of measures. Much of the treatment of operators on $C(K)$ comes from the wonderful book of Dunford and Schwartz [12], which continues to inspire over forty years since its publication. The proof of the Principle of Local Reflexivity given in the text is adapted from that of Dean [5]. For reasons of space, we have not included ultraproducts of Banach spaces, which are needed to make full use of this principle and of the local methods employed in the text. The survey article of Heinrich [24] is an excellent introduction to this invaluable tool.

Chapter 6. The paper [45] by Saphar is recommended. Full treatments of the Chevet-Saphar tensor norms can be found in the Gilbert–Leih paper [17] and the book of Defant and Floret [6]. The class of p-summing operators was introduced by Pietsch [38]. This paper is clearly written and is recommended reading. For p-summing operators and all related matters, there is no better reference than the book of Diestel, Jarchow and Tonge [9]. The proof of Grothendieck's Inequality presented in the text is that of Kaijser and is adapted from the version given in this book.

Chapter 7. We have drawn heavily on the paper of Gilbert and Leih [17] and the book of Defant and Floret [6]. This latter work contains virtually everything that one might wish to learn about tensor norms and is highly recommended to the reader. For some early papers on the p-integral operators, see [36, 37, 39]. The proof of the Kwapien Domination Theorem is taken from the very accessible original [29].

Chapter 8. The paper of Gordon, Lewis and Retherford [18] is very readable and illustrates the principle that tensor product and operator ideal methods are often both needed. The books of Pietsch [40] and Defant-Floret [6] contain complete accounts of the theory of operator ideals. Defant and Floret give an exhaustive development of the interconnections between operator ideals and tensor norms.

Appendix B. For a highly readable account of sequences and series in Banach spaces, see the book by Diestel [8].

Appendix C. For this material, we have drawn on the books of Diestel [8] and Dunford–Schwartz [12].

B. Summability in Banach Spaces

A sequence (x_n) in a Banach space is said to be *summable* if the series $\sum_n x_n$ converges, or, in other words, the partial sums $\sum_{j=1}^{n} x_j$ form a Cauchy sequence. The sequence (x_n) is *absolutely summable* if the series $\sum_n \|x_n\|$ converges. It is a straightforward consequence of the completeness of X that every absolutely summable sequence in X is summable. We now consider some other forms of summability.

We say that (x_n) is *unconditionally summable* if for every permutation σ of the set of natural numbers, the permuted sequence $(x_{\sigma(n)})$ is summable.

The sequence (x_n) is *subseries summable* if, for every strictly increasing sequence (n_k) of natural numbers, the sequence (x_{n_k}) is summable.

The sequence (x_n) is *sign summable* if the sequence $(\varepsilon_n x_n)$ is summable for every choice of signs $\varepsilon_n = \pm 1$.

Finally, we have a condition involving convergence of the net of finite partial sums. Let $\mathfrak{F}$ be the set of finite subsets of $\mathbb{N}$. Then $\mathfrak{F}$ is directed by inclusion and we may define a net $(S_A)_{A\in\mathfrak{F}}$ by $S_A = \sum_{n\in A} x_n$. We say that the sequence (x_n) is *unordered summable* if this net converges. Thus, (x_n) is unordered summable if and only if there exists $x \in X$ with the following property: for every $\varepsilon > 0$, there is a finite subset A of $\mathbb{N}$ such that if B is any finite subset of $\mathbb{N}$ that contains A, then

$$\left\| x - \sum_{n\in B} x_n \right\| < \varepsilon .$$

Equivalently, the net (S_A) satisfies the Cauchy condition: for every $\varepsilon > 0$, there exists a finite subset A of $\mathbb{N}$ such that if B is any finite subset of $\mathbb{N}$ that is disjoint from A, then $\|\sum_{n\in B} x_n\| < \varepsilon$.

Proposition B.1. *The following statements are equivalent for a sequence (x_n) in a Banach space:*

(i) *(x_n) is unconditionally summable.*
(ii) *(x_n) is subseries summable.*
(iii) *(x_n) is sign summable.*
(iv) *(x_n) is unordered summable.*

Proof. (i) implies (ii): we prove the contrapositive. Suppose that (x_n) is not subseries summable, so that some subsequence (x_{n_k}) is not summable.

Therefore there exists $\varepsilon > 0$ such that, for every natural number k, there exist natural numbers l and m with the property that $k < l < m$ and

$$\left\|\sum_{j=l}^{m} x_{n_j}\right\| \geq \varepsilon .$$

Hence there exist mutually disjoint finite subsets I_m of $\mathbb{N}$ such that

$$\left\|\sum_{k\in I_m} x_{n_k}\right\| \geq \varepsilon$$

for every m. Taking a permutation of $\mathbb{N}$ that maps the sets I_m into blocks of consecutive natural numbers yields a rearrangement of the sequence (x_n) whose partial sums do not satisfy the Cauchy condition. Therefore (x_n) is not unconditionally summable.

(ii) implies (iii): suppose that (x_n) is subseries summable and let $\varepsilon_n = \pm 1$ for every n. Let $A = \{n \in \mathbb{N} : \varepsilon_n = 1\}$ and $B = \{n \in \mathbb{N} : \varepsilon_n = -1\}$. Then the series $\sum_{n\in A} x_n$ and $\sum_{n\in B}(-x_n)$ both converge. Since the series $\sum_n \varepsilon_n x_n$ is obtained by interlacing these series, it follows that this series is also convergent. Therefore (x_n) is sign summable.

(iii) implies (iv): suppose that (x_n) is not unordered summable. Then there exists $\varepsilon > 0$ and a sequence of finite subsets A_k of $\mathbb{N}$ such that $\max A_k < \min A_{k+1}$ for every k and

$$\left\|\sum_{n\in A_k} x_n\right\| \geq \varepsilon$$

for every k. Let

$$\varepsilon_n = \begin{cases} 1 & \text{if } n \in \bigcup_k A_k , \\ -1 & \text{otherwise} . \end{cases}$$

Then the series $\sum_n (1 + \varepsilon_n) x_n$ does not satisfy the Cauchy condition. But this series is the sum of the series $\sum_n x_n$ and $\sum_n \varepsilon_n x_n$ and so at least one of these series diverges. Therefore (x_n) is not sign summable.

(iv) implies (i): suppose that (x_n) is unordered summable. Let σ be a permutation of $\mathbb{N}$ and let $\varepsilon > 0$. There exists a finite subset A of $\mathbb{N}$ such that $\|\sum_{n\in B} x_n\| < \varepsilon$ for every finite set B that is disjoint from B. Choose $k \in \mathbb{N}$ such that A is contained in the set $\{\sigma(n) : 1 \leq n \leq k\}$. Then $\|\sum_{j=n}^{m} x_j\| < \varepsilon$ if $k < n < m$. Therefore the series $\sum_n x_{\sigma(n)}$ satisfies the Cauchy condition. □

The strong summability condition expressed in this result is stronger than summability; indeed, there are scalar sequences that are summable but not absolutely summable and it follows easily that the same is true in every Banach space.

In finite dimensions, unconditional and absolute summability are the same. However, the Dvoretzky–Rogers Theorem states that every infinite dimensional Banach space contains an unconditionally summable sequence that is not absolutely summable (see Corollary 6.21 for a weak form of this theorem).

Apart from absolute summability, all the modes of summability defined above have weak counterparts, where convergence with respect to the norm topology is replaced by weak convergence. The weak version of the proposition is false. For example, consider the sequence (x_n) in c_0 given by $x_1 = e_1$ and $x_n = e_n - e_{n-1}$ for all $n > 1$. It is easy to see that this sequence is weakly unordered summable, with sum zero. However, (x_n) is clearly not weakly subseries summable.

Proposition B.2. *Let (x_n) be a sequence in a Banach space.*

(a) *(x_n) is weakly unconditionally summable if and only if it is weakly unordered summable.*
(b) *(x_n) is weakly subseries summable if and only if it is weakly sign summable.*
(c) *If (x_n) is weakly subseries summable then it is weakly unconditionally summable.*

Proof. (a) Suppose that (x_n) is weakly unconditionally summable. Let x be the sum of the weakly convergent series $\sum_n x_n$. Then $\varphi(x)$ is the sum of the unconditionally convergent scalar series $\sum_n \varphi(x_n)$ for every $\varphi \in X^*$ and it follows that the value of the weak sum $\sum_n x_n$ is independent of the ordering of the terms. Now, weak unconditional summability of (x_n) means that the series $\sum_n \varphi(x_n)$ is unconditionally convergent to $\varphi(x)$ for every $\varphi \in X^*$. On the other hand, weak unordered summability of (x_n) means that there exists $z \in X^*$ such that the net of finite partial sums of the series $\sum_n \varphi(x_n)$ converges to $\varphi(z)$ for every $\varphi \in X^*$. The result now follows from the equivalence of these definitions of convergence for scalar series.

(b) The proof of the fact that subseries summability implies sign summability in the previous proposition also applies to the corresponding weak notions. Conversely, suppose that (x_n) is weakly sign summable and let (x_{n_k}) be a subsequence. Let ε_n be 1 if n lies in the range of the sequence (n_k); otherwise, let $\varepsilon_n = -1$. Then the series $\sum_n x_n$ and $\sum_n \varepsilon_n x_n$ are both weakly convergent. Adding, we see that the subseries $\sum_k x_{n_k}$ is also weakly convergent.

(c) Suppose that (x_n) is weakly subseries summable. Then (x_n) is weakly sign summable, by (b). Let x be the sum of the weakly convergent series $\sum_n x_n$ and let $\varphi \in X^*$. Then $\sum_n \varphi(x_n)$ converges to $\varphi(x)$. But the sign summability of (x_n) implies that the series $\sum_n \varepsilon_n \varphi(x_n)$ converges for every choice of signs ε_n. Therefore the scalar series $\sum_n \varphi(x_n)$ is unconditionally convergent to $\varphi(x)$. Since this holds for every $\varphi \in X^*$, it follows that (x_n) is weakly unconditionally summable. □

This result is not the whole story. The Orlicz–Pettis Theorem (Proposition 3.12) asserts that weak subseries summability is equivalent to unconditional summability.

We now present a useful characterization of unconditional summability.

Proposition B.3 (Bounded Multiplier Test). *A sequence (x_n) in a Banach space is unconditionally summable if and only if the sequence $(\alpha_n x_n)$ is summable for every bounded sequence of scalars (α_n).*

Proof. Suppose that $(\alpha_n x_n)$ is summable for every bounded scalar sequence (α_n). Then (x_n) is sign summable and hence is absolutely summable.

Conversely, suppose that (x_n) is unconditionally summable. Let (α_n) be a bounded sequence of scalars. We shall prove that the partial sums of the series $\sum_n \alpha_n x_n$ form a Cauchy sequence. First, we have

$$\left\|\sum_{k=n}^{m} \alpha_k x_k\right\| = \sup_{\varphi\in B_{X^*}} \left|\sum_{k=n}^{m} \alpha_k \varphi(x_k)\right| \le \|(\alpha_n)\|_\infty \sup_{\varphi\in B_{X^*}} \sum_{k=n}^{m} |\varphi(x_k)|$$

for $n < m$. In the complex case, we use the inequality $|\varphi(x)| \le |\operatorname{Re}\varphi(x)| + |\operatorname{Im}\varphi(x)|$ and apply the argument below to the real linear functionals $\operatorname{Re}\varphi$ and $\operatorname{Im}\varphi$.

Proceeding with the real case, let A consist of the natural numbers k between n and m for which $\varphi(x_k) \ge 0$ and let B consist of those k between n and m for which $\varphi(x_k) < 0$. Then

$$\begin{aligned}\sum_{k=n}^{m} |\varphi(x_k)| &= \sum_{k\in A} \varphi(x_k) - \sum_{k\in B} \varphi(x_k)\\ &= \varphi\Big(\sum_{k\in A} x_k\Big) - \varphi\Big(\sum_{k\in B} x_k\Big) \le \Big\|\sum_{k\in A} x_k\Big\| + \Big\|\sum_{k\in B} x_k\Big\|\end{aligned}$$

for every $\varphi \in B_{X^*}$. Since (x_n) is unordered summable, it follows that the series $\sum_n \alpha_n x_n$ satisfies the Cauchy condition. □

We conclude with a weak version of the Bounded Multiplier Test. A series $\sum_n x_n$ is said to be *weakly unconditionally Cauchy* if the partial sums of the series $\sum_n x_{\sigma(n)}$ form a weak Cauchy sequence for every permutation σ of $\mathbb{N}$. Equivalently, the scalar series $\sum_n \varphi(x_n)$ is unconditionally convergent for every $\varphi \in X^*$.

Proposition B.4. *Let (x_n) be a sequence in a Banach space X. The scalar sequence $(\varphi(x_n))$ is unconditionally summable for every $\varphi \in X^*$ if and only if the series $\sum_n \alpha_n x_n$ converges for every $(\alpha_n) \in c_0$.*

Proof. Suppose that $\sum_n |\varphi(x_n)| < \infty$ for every $\varphi \in X^*$. Then we may define a linear mapping $T\colon X^* \to \ell_1$ by $T(\varphi) = (\varphi(x_n))$. An application of the Closed Graph Theorem shows that T is bounded. Therefore

$$\sup\left\{\sum_n |\varphi(x_n)| : \varphi \in B_{X^*}\right\} \leq \|T\| < \infty\,.$$

Let $(\alpha_n) \in c_0$. Arguing as in the proof of the previous proposition, we have

$$\left\|\sum_{k=n}^{m} \alpha_k x_k\right\| \leq \sup_{k\geq n} |\alpha_k|\, \|T\|\,,$$

which tends to zero with n. It follows that the series $\sum_n \alpha_n x_n$ satisfies the Cauchy condition.

Conversely, suppose that $\sum_n \alpha_n x_n$ converges for every $(\alpha_n) \in c_0$. Then the series $\sum_n \alpha_n \varphi(x_n)$ converges for every $(\alpha_n) \in c_0$ and every $\varphi \in X^*$. Hence the sequence $(\varphi(x_n))$ is unconditionally summable for every $\varphi \in X^*$. □

C. Spaces of Measures

In this appendix, we give an account of some of the structural properties of Banach spaces of scalar measures that we need. We are particularly interested in questions relating to the weak topology. We shall assume that the reader has a basic knowledge of measure theory up to and including the Radon–Nikodým Theorem.

We begin by recalling some basic results. Let Σ be a σ-algebra of subsets of a set Ω. We shall refer to the elements of Ω simply as *measurable sets.* We denote by $\mathcal{M}(\Sigma)$ the vector space of scalar valued measures on Σ, that is, functions from Ω into $\mathbb{K} = \mathbb{R}$ or $\mathbb{C}$ that are countably additive and send $\emptyset$ to zero. We emphasize that our measures take only *finite* values. Nevertheless, in order to avoid any confusion, we shall refer to a measure that takes its values in $(0, \infty)$ as a *finite, positive measure.*

The *variation* of a measure μ is defined on Σ by

$$|\mu|(E) = \sup\left\{\sum_{i=1}^{n} |\mu(A_i)| : \{A_1, \dots, A_n\} \text{ a partition of } E\right\},$$

where a *partition* of E is a finite collection of mutually disjoint measurable sets whose union is E. The variation of μ is a finite, positive measure on Σ; it can be characterized as the smallest finite, positive measure that satisfies $|\mu(E)| \leq |\mu|(E)$ for every $E \in \Sigma$. Furthermore, we have

$$|\mu(E)| \leq |\mu|(E) \leq 4 \sup\{|\mu(F)| : F \subset E,\ F \in \Sigma\}. \tag{C.1}$$

In particular, every scalar measure is bounded. The *variation norm* is defined on the vector space $\mathcal{M}(\Sigma)$ by

$$\|\mu\| = |\mu|(\Omega).$$

We will see shortly that $\mathcal{M}(\Sigma)$ is complete in the variation norm.

Let λ be a finite, positive measure on Σ. The measure μ is *absolutely continuous* with respect to λ if, for every $\varepsilon > 0$ there exists $\delta > 0$ such that if $\lambda(E) < \delta$, then $|\mu(E)| < \varepsilon$. Equivalently, if $\lambda(E) = 0$, then $\mu(E) = 0$. We also say that λ is a *control measure* for μ. The Radon–Nikodým Theorem states that μ is absolutely continuous with respect to λ if and only if there exists $f \in L_1(\lambda)$ such that $\mu(E) = \int_E f\, d\lambda$ for every measurable set E.

Furthermore, we have $|\mu|(E) = \int_E |f|\,d\lambda$ and it follows in particular that the variation norm of μ coincides with the $L_1(\lambda)$-norm of f.

For each finite, positive measure λ, we denote by $\mathcal{M}_\lambda(\Sigma)$ the subspace of $\mathcal{M}(\Sigma)$ consisting of all the measures that are absolutely continuous with respect to λ. The remarks above show that the correspondence between a measure and its Radon–Nikodým derivative establishes an isometric isomorphism between $\mathcal{M}_\lambda(\Sigma)$ and $L_1(\lambda)$. This observation is invaluable when investigating properties of $\mathcal{M}(\Sigma)$ relating to the behaviour of sequences of measures. Indeed, if (μ_n) is any sequence of measures on Σ, we may define a finite, positive measure by

$$\lambda(E) = \sum_{n=1}^{\infty} 2^{-n}(1 + \|\mu_n\|)^{-1}|\mu_n|(E)\,.$$

It is easy to see that if $\lambda(E) = 0$, then $\mu_n(E) = 0$ for every n. Thus λ is a control measure for all the measures μ_n. To summarize, we have shown that every sequence in $\mathcal{M}(\Sigma)$ lies in a subspace that is isometrically isomorphic to some $L_1(\lambda)$. We now give some applications of this idea.

Proposition C.1. *The space $\mathcal{M}(\Sigma)$ is complete in the variation norm.*

Proof. Let (μ_n) be a Cauchy sequence in $\mathcal{M}(\Sigma)$. This sequence lies in a subspace of $\mathcal{M}(\Sigma)$ that is isomorphic to the Banach space $L_1(\lambda)$ for some λ. Therefore (μ_n) converges. □

Next, we have a characterization of weak convergence:

Proposition C.2. *A bounded sequence (μ_n) in $\mathcal{M}(\Sigma)$ converges weakly to μ if and only if $(\mu_n(E))$ converges to $\mu(E)$ for every $E \in \Sigma$.*

Proof. The necessity of the stated contition follows from the fact that the mapping $E \mapsto \mu(E)$ defines a bounded linear functional on $\mathcal{M}(\Sigma)$ for every $E \in \Sigma$.

To prove sufficiency, it suffices to deal with the case $\mu = 0$. Arguing as in the proof of the preceding proposition, the sequence (μ_n) lies in a subspace $\mathcal{M}_\lambda(\Sigma)$ that is isometrically isomorphic to $L_1(\lambda)$, so that $\mu_n(E) = \int_E f_n\,d\lambda$ for every $E \in \Sigma$ and every n, where $f_n \in L_1(\lambda)$. Then (f_n) is a bounded sequence in $L_1(\lambda)$ and the condition $\mu_n(E) \to 0$ is equivalent to $\int_E f_n\,d\lambda \to 0$ for every E.

We wish to show that $\int_E fg\,d\lambda \to 0$ for every $g \in L_\infty(\lambda)$. By the boundedness of the sequence (f_n) and the density of the measurable simple functions in $L_\infty(\lambda)$, it is enough to show that $\int_\Omega g f_n\,d\lambda \to 0$ for every measurable simple function g. This is easily seen to be equivalent to the condition that $\int_E f_n\,d\lambda \to 0$ for every measurable set E. Therefore (f_n) converges weakly to zero in $L_1(\lambda)$ and hence (μ_n) converges weakly to zero. □

Now fix a finite, positive measure λ. For each measurable set, E, the characteristic function χ_E is integrable with respect to λ. Two characteristic functions, χ_E and χ_F, belong to the same equivalence class in $L_1(\lambda)$ if they are equal almost everywhere with respect to λ or, equivalently, if the symmetric difference $E \triangle F$ has λ-measure zero. This defines an equivalence relation on Σ. We denote the quotient set by Σ_λ and for each $E \in \Sigma$, we denote its equivalence class by $\overline{E}$. We may now consider Σ_λ as a subset of $L_1(\lambda)$ via the injective mapping $\overline{E} \mapsto \chi_E$. This embedding induces a metric on Σ_λ:

$$d_\lambda(\overline{E}, \overline{F}) = \int_\Omega |\chi_E - \chi_F| \, d\lambda = \lambda(E \triangle F).$$

We claim that the metric space $(\Sigma_\lambda, d_\lambda)$ is complete. To see this, let $(\overline{E}_n)$ be a Cauchy sequence. Then the sequence (χ_{E_n}) is Cauchy in $L_1(\lambda)$ and so converges. Let the function f be a representative of the limit of this sequence. Since every L_1-convergent sequence has a subsequence that converges almost everywhere, it follows that f takes the values 0, 1 almost everywhere with respect to λ. Therefore f lies in the same equivalence class as the characteristic function of some measurable set E. It follows that the sequence $(\overline{E}_n)$ converges to $\overline{E}$.

The fact that $(\Sigma_\lambda, d_\lambda)$ is a complete metric space opens the door to the Baire Category Theorem. All we need to add to the picture is the fact that absolute continuity is the same as continuity in the metric sense. Suppose that μ is absolutely continuous with respect to λ. Then μ is a well-defined function on Σ_λ, since $\lambda(E \triangle F) = 0$ implies $\mu(E \triangle F) = 0$ and hence $\mu(E) = \mu(F)$. Then the $\varepsilon - \delta$ definition of absolute continuity shows that μ is a continuous function on the metric space $(\Sigma_\lambda, d_\lambda)$.

Proposition C.3 (Vitali–Hahn–Saks Theorem). *Let (μ_n) be a sequence of measures on the σ-algebra Σ such that $(\mu_n(E))$ converges for every $E \in \Sigma$, and suppose that each μ_n is absolutely continuous with respect to the same finite, positive measure λ. Then μ_n are uniformly absolutely continuous with respect to λ and the formula $\mu(E) = \lim_n \mu_n(E)$ defines a measure on Σ that is also absolutely continuous with respect to λ.*

Proof. By the remarks preceding the proposition, we may consider the measures μ_n as continuous functions on the metric space $(\Sigma_\lambda, d_\lambda)$. Fix $\varepsilon > 0$ and consider the sets

$$F_k = \left\{ \overline{A} \in \Sigma_\lambda : |\mu_n(A) - \mu_m(A)| \le \varepsilon \quad \forall m, n \ge k \right\}.$$

The continuity of the μ_n ensures that these sets are closed and the fact that the sequence $(\mu_n(A))$ converges for every $A \in \Sigma$ implies that the union of the F_k is all of Σ_λ. Hence, by the Baire Category Theorem, there exists k such that F_k has non-empty interior. Thus there exists $A_0 \in \Sigma$ and $\delta > 0$ such that

$$\lambda(E \triangle A_0) < \delta \Rightarrow |\mu_n(E) - \mu_m(E)| \le \varepsilon \quad \forall m, n \ge k.$$

Let E be a measurable set such that $\lambda(E) < \delta$. We can write E as $E_1 \setminus E_2$, where $E_1 = E \cup A_0$ and $E_2 = A_0 \setminus E$. Then $\lambda(E_i \triangle A_0) < \delta$ for $i = 1, 2$. Therefore, if $n \ge k$, we have

$$\begin{aligned} |\mu_n(E)| &\le |\mu_k(E)| + |\mu_n(E) - \mu_k(E)| \\ &= |\mu_k(E)| + |\mu_n(E_1) - \mu_n(E_2) - \mu_k(E_1) + \mu_k(E_2)| \\ &\le |\mu_k(E)| + |\mu_n(E_1) - \mu_k(E_1)| + |\mu_n(E_2) - \mu_k(E_2)| \\ &\le |\mu_k(E)| + 2\varepsilon. \end{aligned}$$

By the absolute continuity of $\mu_1, \ldots, \mu_k$, we may assume that $\lambda(E) < \delta$ implies $|\mu_i(E)| < \varepsilon$ for $1 \le i \le k$. Hence we have shown that if $\lambda(E) < \delta$, then $|\mu_n(E)| \le 3\varepsilon$ for every n. This proves the uniform absolute continuity of the sequence (μ_n).

Now consider the function $\mu(E) = \lim_n \mu_n(E)$. It is easy to see that μ is finitely additive and so, in order to establish the countable additivity of μ, it suffices to show that $\mu(E_k) \to 0$ for every decreasing sequence (E_k) of measurable sets with empty intersection. But this follows immediately from the uniform absolute continuity of the μ_n, as does the absolute continuity of μ. □

The presence of the measure λ here is something of a red herring, as we have seen that every sequence of measures has a control measure. We may rephrase the result without explicit mention of λ:

Proposition C.4 (Nikodým Convergence Theorem). *Let (μ_n) be a sequence of measures on Σ such that the sequence $(\mu_n(E))$ converges for every $E \in \Sigma$. Then the μ_n are uniformly countably additive and $\mu(E) = \lim_n \mu_n(E)$ defines a measure on Σ.*

Combining the Nikodým Convergence Theorem with the characterization of weak convergence given in Proposition C.2, we obtain:

Corollary C.5. *The Banach space $\mathcal{M}(\Sigma)$ is weakly sequentially complete.*

As a byproduct of this result, we see that the space $L_1(\lambda)$ is weakly sequentially complete, since this space is a subspace of $\mathcal{M}(\Sigma)$.

We can also deduce a striking property of the space ℓ_1. If Ω is the set of natural numbers and Σ the power set, then $\mathcal{M}(\Sigma)$ is ℓ_1, where the measure μ is given by its values $\mu_k = \mu(\{k\})$ at the singletons and the variation of μ on a set E is given by $\sum_{k \in E} |\mu_k|$. A Banach space is said to have the *Schur property* if every weakly convergent sequence converges in norm.

Corollary C.6. *The Banach space ℓ_1 has the Schur property.*

Proof. Suppose the sequence (μ_n) converges weakly to zero in ℓ_1, where $\mu_n = (\mu_{nk})_k$ for each n. By Proposition C.3, there exists a positive element, λ, of ℓ_1 with respect to which the μ_n are uniformly absolutely continuous. Let $\varepsilon > 0$. There exists $\delta > 0$ such that, if $\sum_{k\in E} \lambda_k < \delta$, then $|\sum_{k\in E} \mu_{nk}| < \varepsilon$ for every n. It follows from (C.1) that, if $\sum_{k\in E} \lambda_k < \delta$, then $\sum_{k\in E} |\mu_{nk}| < 4\varepsilon$ for every n. Choose a positive integer K so that $\sum_{k>K} \lambda_k < \delta$. Then $\sum_{k>K} |\mu_{nk}| < 4\varepsilon$ for every n. Since (μ_n) is weakly convergent to zero, there is a positive integer N such that $\sum_{k=1}^{K} |\mu_{nk}| < \varepsilon$ if $n \geq N$. Therefore

$$\|\mu_n\| = \sum_{k=1}^{K} |\mu_{nk}| + \sum_{k>K} |\mu_{nk}| < 5\varepsilon$$

when $n \geq N$. Hence (μ_n) converges to zero in norm. □

We conclude with a characterization of the weakly compact sets in $\mathcal{M}(\Sigma)$. We say that the elements of a subset, K, of $\mathcal{M}(\Sigma)$ are *uniformly countable additive* if, for every sequence (E_n) of mutually disjoint measurable sets, the convergence of the series $\sum_n \mu(E_n)$ is uniform in $\mu \in K$. Equivalently, if (A_n) is a decreasing sequence of measurable sets with empty intersection, then the sequence $(\mu(A_n))$ converges to zero uniformly in $\mu \in K$.

Proposition C.7. *The following statements are equivalent for a bounded subset K of $\mathcal{M}(\Sigma)$:*

(i) *K is relatively weakly compact.*
(ii) *There exists a finite, positive measure λ on Σ such that the elements of K are uniformly absolutely continuous with respect to λ.*
(iii) *The elements of K are uniformly countably additive.*

Proof. (i) implies (ii): suppose that K is relatively weakly compact. We claim that for every $\varepsilon > 0$ there exists $\delta > 0$ and a finite subset $\{\mu_1, \ldots, \mu_m\}$ of K such that if $|\mu_j|(E) < \delta$ for $1 \leq j \leq m$, then $|\mu(E)| < \varepsilon$ for every $\mu \in K$.

Suppose for the moment this is true. Taking $\varepsilon = 2^{-n}$, we get $\delta_n > 0$ and elements μ_{nj} of K, for $1 \leq j \leq m_n$, such that for every n,

$$\max_{1\leq j\leq m_n} |\mu_{nj}|(E) < \delta \quad \Rightarrow \quad |\mu(E)| < 2^{-n} \quad \text{for every } \mu \in K. \qquad (*)$$

Let

$$\nu_n = \frac{1}{m_n} \sum_{j=1}^{m_n} (1 + \|\mu_{nj}\|)^{-1} |\mu_{nj}|.$$

Then $\|\nu_n\| \leq 1$ for every n and so

$$\lambda = \sum_{n=1}^{\infty} 2^{-n} \nu_n$$

defines a finite, positive measure on Σ. Now suppose that

$$\lambda(E) < \frac{2^{-n}\delta_n}{m_n(1+M)},$$

where $M = \sup_{\mu \in K} \|\mu\|$. Then

$$\nu_n(E) < \delta_n / m_n(1+M)$$

and it follows that

$$\frac{|\mu_{nj}|(E)}{1+\|\mu_{nj}\|} \leq m_n \nu_n(E) < \frac{\delta_n}{1+M}$$

and so $|(\mu_{nj}|(E) < \delta_n$ for each $j = 1, \ldots, m_n$. Hence, by $(*)$, we have $|\mu(E)| < 2^{-n}$ for every $\mu \in K$. This establishes the uniform absolute continuity of the elements of K with respect to λ.

We prove our claim by contradiction. Suppose it is false for some $\varepsilon > 0$. Taking $\delta = 2^{-1}$ and fixing any $\mu_1 \in K$, there exists $E_1 \in \Sigma$ and $\mu_2 \in K$ such that $|\mu_1|(E_1) < 2^{-1}$ and $|\mu_2(E_1)| \geq \varepsilon$. Similarly, taking $\delta = 2^{-2}$, there exists $E_2 \in \Sigma$ and $\mu_3 \in K$ such that

$$|\mu_1|(E_2), |\mu_2|(E_2) < 2^{-2} \quad \text{and} \quad |\mu_3(E_2)| \geq \varepsilon .$$

Proceeding in this way, we get a sequence (μ_n) in K and a sequence (E_n) of measurable sets such that

$$|\mu_j|(E_n) < 2^{-n} \quad \text{for } 1 \leq j \leq n \text{ and } \quad |\mu_{n+1}(E_n)| \geq \varepsilon$$

for every n. By the relative weak compactness of K, we may assume that the sequence (μ_n) is weakly convergent. Consider the control measure for this sequence given by

$$\nu(E) = \sum_{n=1}^{\infty} \frac{2^{-n}|\mu_n|(E)}{1+\|\mu_n\|}.$$

By the Vitali–Hahn–Saks Theorem, the μ_n are uniformly absolutely continuous with respect to ν. However, we have

$$\nu(E_n) \leq \sum_{k=1}^{n} \frac{2^{-k}|\mu_k|(E_n)}{1+\|\mu_k\|} + \sum_{k=n+1}^{\infty} 2^{-k}$$
$$< \sum_{k=1}^{n} 2^{-k}2^{-n} + \sum_{k=n+1}^{\infty} 2^{-k} \leq 2^{-n+1}$$

for every n, while $|\mu_{n+1}(E_n)| \geq \varepsilon$, a contradiction. Thus our claim is proved.

(ii) implies (iii) is trivial.

(iii) implies (i): suppose that the elements of K are uniformly countably additive. Let (μ_n) be a sequence in K. Choose a control measure λ for this sequence. Then this sequence lies in the subspace $\mathcal{M}_\lambda(\Sigma)$ of $\mathcal{M}(\Sigma)$ that is isometrically isomorphic to $L_1(\lambda)$, each measure μ_n corresponding to its Radon–Nikodým derivative f_n. Taking a countable base $\{O_m\}$ for the topology of $\mathbb{K}$, let $E_{nm} = f_n^{-1}(O_m)$. Let Σ_1 be the sub-σ-algebra of Σ generated by the countable family of sets $\{E_{nm}\}$ and let λ_1 be the restriction of the measure λ to Σ_1. Then the sequence (f_n) lies in the closed subspace $L_1(\lambda_1)$ of $L_1(\lambda)$. We shall show that the sequence (f_n) has a subsequence that converges weakly in $L_1(\lambda_1)$.

Since the sequence $(\mu_n(E))$ is bounded for every measurable set E, a diagonal argument shows that (μ_n) contains a subsequence (μ_{n_k}) with the property that the sequence $(\mu_{n_k}(E_{nm}))$ converges for each member of the countable family of sets $\{E_{nm}\}$. Let Γ be the collection of all sets $E \in \Sigma_1$ for which the sequence $(\mu_{n_k}(E))$ converges. Thus, every set in the algebra generated by the sets E_{nm} belongs to Γ. We claim that Γ is a monotone class. To see this, let (F_j) be an increasing sequence in Γ and let $F = \bigcup_j F_j$. By the uniform countable additivity of the sequence (μ_n), we have $\mu_{n_k}(F_j) \to \mu_{n_k}(F)$ uniformly in k. Furthermore, the sequence $(\mu_{n_k}(F_j))_k$ converges for each j. These facts together imply that the sequence $(\mu_{n_k}(F))$ converges and hence F belongs to Σ. A similar argument shows that Γ is closed under countable decreasing intersections.

Therefore Γ is a monotone class that contains the algebra generated by the sets E_{nm}. By the Monotone Class Lemma, Γ contains the σ-algebra generated by the E_{nm} and so the sequence $(\mu_{n_k}(E))$ converges for every $E \in \Sigma_1$. Since the sequence (μ_{n_k}) is bounded, it follows that (f_{n_k}) converges weakly in $L_1(\lambda_1)$ and hence also in $L_1(\lambda)$. Therefore (μ_{n_k}) converges weakly in $\mathcal{M}(\Sigma)$. □

Since the spaces $L_1(\lambda)$ are subspaces of $\mathcal{M}(\Sigma)$, we also obtain a description of the weakly compact subsets of these spaces. A subset K of $L_1(\lambda)$ is said to be *uniformly integrable* if, for every $\varepsilon > 0$, there exists $\delta > 0$ such that if $\lambda(E) < \delta$, then $\int_E |f|\, d\lambda < \varepsilon$ for every $f \in K$. It is easy to see that this notion is equivalent to uniform countable additivity. Thus, we have:

Corollary C.8. *Let λ be a finite, positive measure. A bounded subset, K, of $L_1(\lambda)$ is relatively weakly compact if and only if the elements of K are uniformly integrable.*

Most of the results presented in this appendix have a "topological version", where the space of measures $\mathcal{M}(\Sigma)$ is replaced by the space of regular Borel measures on a compact topological space. We shall need just one of these topological results in the text. Let S be a compact topological space and let $\mathcal{B}_S$ be the σ-algebra of Borel subsets. Consider the Banach space $C(S)$ of continuous scalar functions on S, endowed with the usual supremum norm. The Riesz Representation Theorem states that the dual space $C(S)^*$ is the space of regular Borel measures on $\mathcal{B}_S$, with the variation norm. We note that $C(S)^*$ is a closed subspace of $\mathcal{M}(\mathcal{B}_S)$. We will need a characterization of the weakly compact subsets of $C(S)^*$. An examination of the proof of Proposition C.7 shows that the control measure constructed in the proof of the implication (i) $\Rightarrow$ (ii) is regular if all the measures belonging to the set K are regular. Thus, we have the following:

Proposition C.9. *Let S be a compact topological space. The following are equivalent for a bounded subset K of $C(S)^*$:*

(i) *K is relatively weakly compact.*
(ii) *There exists a finite, positive, regular measure λ on $\mathcal{B}_S$ such that the elements of K are uniformly absolutely continuous with respect to λ.*
(iii) *The elements of K are uniformly countably additive.*

References

1. I. Amemiya and K. Shiga. On tensor products of Banach spaces. *Kōdai Math. Sem. Rep.*, 9:161–178, 1957.
2. R. D. Bourgin. *Geometric aspects of convex sets with the Radon-Nikodým property.* Springer-Verlag, Berlin, 1983.
3. S. Chevet. Sur certains produits tensoriels topologiques d'espaces de Banach. *Z. Wahrscheinlichkeitstheorie und Verw. Gebiete*, 11:120–138, 1969.
4. Simone Chevet. Sur certains produits tensoriels topologiques d'espaces de Banach. *C. R. Acad. Sci. Paris Sér. A-B*, 266:A413–A415, 1968.
5. D. W. Dean. The equation $L(E, X^{**}) = L(E, X)^{**}$ and the principle of local reflexivity. *Proc. Amer. Math. Soc.*, 40:146–148, 1973.
6. A. Defant and K. Floret. *Tensor norms and operator ideals.* North-Holland Publishing Co., Amsterdam, 1993.
7. J. Diestel. The Radon-Nikodým property and spaces of operators. In *Measure theory (Proc. Conf., Oberwolfach, 1975)*, pages 211–227. Lecture Notes in Math., Vol. 541. Springer, Berlin, 1976.
8. J. Diestel. *Sequences and series in Banach spaces.* Springer-Verlag, New York, 1984.
9. J. Diestel, H. Jarchow, and A. Tonge. *Absolutely summing operators.* Cambridge University Press, Cambridge, 1995.
10. J. Diestel and J. J. Uhl, Jr. *Vector measures.* American Mathematical Society, Providence, R.I., 1977. With a foreword by B. J. Pettis, Mathematical Surveys, No. 15.
11. J. Diestel and J. Jerry Uhl, Jr. Progress in vector measures—1977–83. In *Measure theory and its applications (Sherbrooke, Que., 1982)*, pages 144–192. Springer, Berlin, 1983.
12. N. Dunford and J. T. Schwartz. *Linear operators. Part I.* John Wiley & Sons Inc., New York, 1988. General theory, With the assistance of William G. Bade and Robert G. Bartle, Reprint of the 1958 original, A Wiley-Interscience Publication.
13. P. Enflo. A counterexample to the approximation problem in Banach spaces. *Acta Math.*, 130:309–317, 1973.
14. B. R. Gelbaum and J. Gil de Lamadrid. Bases of tensor products of Banach spaces. *Pacific J. Math.*, 11:1281–1286, 1961.
15. J. Gil de Lamadrid. Measures and tensors. *Trans. Amer. Math. Soc.*, 114:98–121, 1965.
16. J. Gil de Lamadrid. Measures and tensors. II. *Canad. J. Math.*, 18:762–793, 1966.

17. J. E. Gilbert and T. J. Leih. Factorization, tensor products, and bilinear forms in Banach space theory. In *Notes in Banach spaces*, pages 182–305. Univ. Texas Press, Austin, Tex., 1980.
18. Y. Gordon, D. R. Lewis, and J. R. Retherford. Banach ideals of operators with applications. *J. Functional Analysis*, 14:85–129, 1973.
19. A. Grothendieck. Résumé de la théorie métrique des produits tensoriels topologiques. *Bol. Soc. Mat. São Paulo*, 8:1–79, 1953.
20. A. Grothendieck. Sur certaines classes de suites dans les espaces de Banach et le théorème de Dvoretzky-Rogers. *Bol. Soc. Mat. São Paulo*, 8:81–110 (1956), 1953.
21. A. Grothendieck. Produits tensoriels topologiques et espaces nucléaires. *Mem. Amer. Math. Soc.*, 1955(16):140, 1955.
22. A. Grothendieck. Résumé de la théorie métrique des produits tensoriels topologiques. *Resenhas*, 2(4):401–480, 1996. Reprint of Bol. Soc. Mat. São Paulo **8** (1953), 1–79 [MR **20** #1194].
23. A. Grothendieck. Sur certaines classes de suites dans les espaces de Banach, et le théorème de Dvoretzky-Rogers. *Resenhas*, 3(4):447–477, 1998. With a foreword by Paulo Cordaro, Reprint of the 1953 original.
24. S. Heinrich. Ultraproducts in Banach space theory. *J. Reine Angew. Math.*, 313:72–104, 1980.
25. J. R. Holub. Tensor product bases and tensor diagonals. *Trans. Amer. Math. Soc.*, 151:563–579, 1970.
26. J. R. Holub. Tensor product mappings. *Math. Ann.*, 188:1–12, 1970.
27. J. R. Holub. Hilbertian operators and reflexive tensor products. *Pacific J. Math.*, 36:185–194, 1971.
28. G. J. O. Jameson. *Summing and nuclear norms in Banach space theory*. Cambridge University Press, Cambridge, 1987.
29. S. Kwapień. On operators factorizable through L_p space. In *Actes du Colloque d'Analyse Fonctionnelle de Bordeaux (Univ. de Bordeaux, 1971)*, pages 215–225. Bull. Soc. Math. France, Mém. No. 31–32. Soc. Math. France, Paris, 1972.
30. S. Kwapień and A. Pełczyński. The main triangle projection in matrix spaces and its applications. *Studia Math.*, 34:43–68, 1970.
31. J.-T. Lapresté. Sur un généralisation de la notion d'opérateurs nucléaires et sommants dans les espaces de banach. *C. R. Acad. Sci. Paris.*, 275:45–48, 1972.
32. J.-T. Lapresté. Opérateurs sommants et factorisations. À travers les espaces L^p. *Studia Math.*, 57(1):47–83, 1976.
33. D. R. Lewis. Conditional weak compactness in certain inductive tensor products. *Math. Ann.*, 201:201–209, 1973.
34. J. Lindenstrauss and A. Pełczyński. Absolutely summing operators in $\mathcal{L}_p$-spaces and their applications. *Studia Math.*, 29:275–326, 1968.
35. J. Lindenstrauss and H. P. Rosenthal. The $\mathcal{L}_p$ spaces. *Israel J. Math.*, 7:325–349, 1969.
36. A. Persson. On some properties of p-nuclear and p-integral operators. *Studia Math.*, 33:213–222, 1969.
37. A. Persson and A. Pietsch. p-nukleare une p-integrale Abbildungen in Banachräumen. *Studia Math.*, 33:19–62, 1969.
38. A. Pietsch. Absolut p-summierende Abbildungen in normierten Räumen. *Studia Mathematica*, 28:333–353, 1967.
39. A. Pietsch. Adjungierte normierte operatorenideale. *Math. Nachr.*, 41:189–211, 1971.

40. A. Pietsch. *Operator ideals*. VEB Deutscher Verlag der Wissenschaften, Berlin, 1978.
41. G. Pisier. Counterexamples to a conjecture of Grothendieck. *Acta Math.*, 151(3-4):181–208, 1983.
42. G. Pisier. *Factorization of linear operators and geometry of Banach spaces*. Published for the Conference Board of the Mathematical Sciences, Washington, DC, 1986.
43. P. Saphar. Comparaison de normes sur des produits tensoriels d'espaces de Banach. Applications. *C. R. Acad. Sci. Paris Sér. A-B*, 266:A809–A811, 1968.
44. P. Saphar. Produits tensoriels topologiques et classes d'applications linéaires. *C. R. Acad. Sci. Paris Sér. A-B*, 266:A526–A528, 1968.
45. P. Saphar. Produits tensoriels d'espaces de Banach et classes d'applications linéaires. *Studia Math.*, 38:71–100. (errata insert), 1970.
46. R. Schatten. *A Theory of Cross-Spaces*. Princeton University Press, Princeton, N. J., 1950.
47. R. Schatten. *Norm ideals of completely continuous operators*. Springer-Verlag, Berlin, 1960.
48. M. Talagrand. Pettis integral and measure theory. *Mem. Amer. Math. Soc.*, 51(307):ix+224, 1984.
49. D. van Dulst. The geometry of Banach spaces with the Radon-Nikodým property. *Rend. Circ. Mat. Palermo (2)*, (Suppl. 7):ix+81, 1985.

Index

MIX
Papier aus verantwortungsvollen Quellen
Paper from responsible sources
FSC® C105338

If you have any concerns about our products,
you can contact us on
ProductSafety@springernature.com

In case Publisher is established outside the EU,
the EU authorized representative is:
Springer Nature Customer Service Center GmbH
Europaplatz 3, 69115 Heidelberg, Germany

Printed by Libri Plureos GmbH
in Hamburg, Germany